MPLS & LABEL SWITCHING NETWORKS

ISBN 0-13-035819-3

Prentice Hall Series in
Advanced Communications Technologies

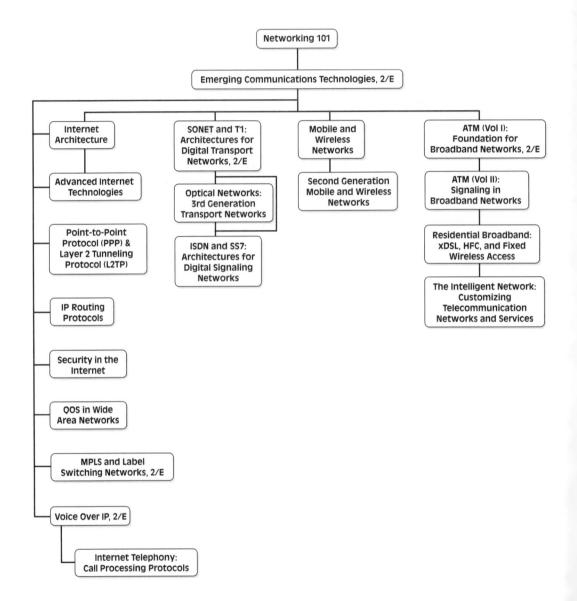

MPLS & LABEL SWITCHING NETWORKS

UYLESS BLACK

PRENTICE HALL SERIES IN ADVANCED
COMMUNICATIONS TECHNOLOGIES

To join a Prentice Hall PTR internet mailing list, point to:
www.phptr.com/register

Prentice Hall PTR
Upper Saddle River, New Jersey 07458
www.phptr.com

Library of Congress Cataloging-in-Publication Data

Black, Uyless D.
   MPLS and label switching networks / Uyless Black.--2nd ed.
     p. cm.--(Prentice Hall series in advanced communications technologies)
   Includes index.
   ISBN 0-13-035819-3
     1. Internet. 2. Computer networks--Management. 3.
  Telecommunications--Traffic--Management. 4. Packet switching (Data transmission) 5.
  MPLS standard. I. Title. II. Series.

   TK5105.875.I57 B55 2002
   004.6'6--dc21

                                           2002018903

Editorial/Production supervision: *Laura Burgess*
Acquisitions editor: *Mary Franz*
Cover designer: *Nina Scuderi*
Cover design director: *Jerry Votta*
Manufacturing manager: *Maura Zaldivar*
Marketing manager: *Dan DePasquale*
Compositor/Production services: *Pine Tree Composition, Inc.*

 © 2002 by Uyless Black
Published by Prentice Hall PTR
A division of Pearson Education, Inc.
Upper Saddle River, New Jersey 07458

Prentice Hall books are widely used by corporations and
government agencies for training, marketing, and resale.

For information regarding corporate and government
bulk discounts please contact:

        Corporate and Government Sales
        Phone: 800-382-3419 or
        corpsales@pearsontechgroup.com

All product names mentioned herein are the trademarks or
registered trademarks of their respective owners.

Printed in the United States of America
10 9 8 7 6 5 4 3 2 1

ISBN: 0-13-035819-3

Pearson Education LTD.
Pearson Education Australia PTY, Limited
Pearson Education Singapore, Pte. Ltd.
Pearson Education North Asia Ltd.
Pearson Education Canada Ltd.
Pearson Educatión de Mexico, S.A. de C.V.
Pearson Education—Japan
Pearson Education Malaysia, Pte. Ltd.

Acknowledgments, Thanks, and Farewell

In my previous books, my dedication prose has been one or two lines of text. This book is an exception, because it is the last book I plan to write, at least about the subject of computer networks. I wish to take this time and space to express my thoughts and thanks to certain people who have been a positive factor in my book-writing endeavors for the past 20 years.

Several of my clients and partners were instrumental to my success. Those cited in this paragraph placed in my hands the responsibility of not only developing an integrated training curriculum on computer network protocols for their companies, but trusted me to be the principal lecturer on all the subjects. It was from their belief in my abilities and their trust in my work that I had the professional and financial support to write 33 books on the subject of computer networks. I give these people my thanks, and I sometimes think they deserve a byline on the covers of my books (however, I am keeping the royalties): Rosemary Aquilar, Satyjit Doctor, Doug Harward, Lisa Henline, Alan Lee, Tanya Mallows, Warren Minami, Ed O'Connor, Jim Opperman, Kathy St. Marie, Herb Stern, and Ken Sherman.

Many people have been a well-spring of support and encouragement, and not just for my books, but for their friendship: Bill and Mary Bacchus, Chock and Patty Black, Marilyn Black, Rob Black, Sandy Black, Lynda Boose, Rich and Judy DeRose, Nicki Deutch, Phil and Marcella Dietz, Bernie and Colette DiTutulillo, Sue Dugan, Pat Fitzpatrick, Frank and Joan Gillen, Al and Carol Hughes, Paul and Trudy Lombardi, Sharon Mahoney, Jeanne Malin, Steve and Ceil Malphrus, Ken and Carol Mare, Ray and Sandy Massey, Karen Nold, Jack and Louise Norris, Gardiner and Pat Pollich, Jack Power, Kathleen Sadler, Jerry and Susan Semrod, Jean Stum, Bob and Betty Taylor, Jo Ellen Thompson, Greg Venit, Patricia Vistins, Beth Waters, Brad and Tahnee Waters, Kathryn Waters, and Sharleen Waters. If a name is not cited in this paragraph, it is likely that the name is on one of my other dedication pages in another book.

My son, my brothers, and their wives have long been supporters of my work: Tommy Black, Ross and Cherrill Black, Ed Black, and Tom and Kaky Black. Don and Peg Black are my cousins, but I think of them as a brother and sister; they belong in this dedication.

My editor at Prentice Hall PTR, Mary Franz, deserves special mention. Like my clients, she also believed in my work, and in my ability to write about many subjects. Many years ago, I told Mary that if she ever left Prentice Hall, I would follow her to the next publishing house. That is the ultimate compliment a writer can give to his editor.

More than anyone else, the person that sustained me through my last 30 book-writing projects was and is my wife Holly. It is a cliché to say that writing can be a lonely profession. Cliché or not, it is a true statement. Holly gave me an extraordinary amount of love and support during this time, and good advice about my books as well. She showed infinite patience during my solitary work. I also think she deserves a by-line on the covers of my books and yes, in case you are curious, she is sharing the well-deserved royalties!

A few of my readers have asked me or my publisher if I had really written all my books, or if I had a ghost writer or a research assistant to "speed things along." The answer is yes. Their names are the ITU-T, the ISO, the EIA, the ANSI, the IEEE, and the IETF, the standards groups that deal with my profession. In many instances, they were the originators of the subjects, and I was the distiller, attempting to bring their (admittedly) rather abstract concepts to the pages my books. My debt to these organizations is immeasurable, as is the debt our society owes to their efforts to build a cohesive framework for our modern telecommunications societies.

It is now time to move to aspects of life beyond bits and bytes. It is possible that I will follow the path of the great football player, John Riggins, who said upon his return from his "retirement": "I'm bored, broke, and back." Yes, it is possible, but not probable. There are other subjects to explore, to occupy my time, and perhaps to write about.

My final thanks go to my readers. After all, it is you who really kept me going through two decades of writing technical books.

If you had not read, I would not have written.

The MPLS technology and standards have matured considerably since the publication of the first edition of this book. Indeed, the first edition was written well before the Internet Engineering Task Force (IETF) approved the MPLS standard as RFC 3031.

This second edition reflects the many changes that have taken place over the past two years, including RFC 3031, as well as many other related topics, such as MPLS-based traffic engineering and MPLS-based optical networks.

The new edition also brings the reader up-to-date on the emerging systems that use revisions to RSVP, BGP, and OSPF to support MPLS networks.

This edition is a different book than the first book. But acting on some readers' requests, I have retained and added basic tutorial material, especially in Chapters 2, 3, and 4.

There are many ways to design and implement an MPLS network. My approach in this book is to show how the Internet standards (and working drafts) published by the IETF can be used to set up and manage label switching routers. I also cite some examples of operations in Cisco routers.

Some of the material pertaining to MPLS interworking with IP and with optical networks was prepared as a result of my research and consulting work with several of my clients, who are producing products and/or services for MPLS networks. I do not claim that my ideas stemming from this work are the only way to build these systems, and I welcome any comments you may have about my approaches.

Contents

CHAPTER 2   Label Switching Basics   **21**

CHAPTER 3   Switching and Forwarding Operations   **38**

CHAPTER 9   Constraint-Based Routing with CR-LDP   211

Preface

As the name of this book implies, the focus is on the switching of traffic though a network or networks. The term *switching* is also known in some parts of the industry as forwarding, relaying, and routing.

INTERNET DRAFTS: WORK IN PROGRESS

In many of my explanations of label switching operations, I have relied on the Internet Request for Comments (RFCs) and draft standards, published by the Internet Society, and I thank this organization for making the RFCs available to the public. The draft standards are "works in progress," and usually change as they wind their way to an RFC (if indeed they become an RFC). A work in progress cannot be considered final, but many vendors use them in creating products for the marketplace. Notwithstanding, they are subject to change.

For all the Internet standards and draft standards the following applies:

1
Introduction

This chapter explains why label switching networks and Multiprotocol Label Switching (MPLS) have become key players in the emerging multiservice public Internet and private internets. It explains the problems associated with conventional IP routing procedures and introduces the concepts of the alternative: label switching. The chapter also introduces the idea of quality of service (QOS) and explains its importance, as well as the importance of label switching to QOS. The chapter concludes with an example of a label switching and QOS network operation at a label switching router (LSR).

WHAT IS LABEL SWITCHING?

The basic concept of label switching is simple. To show why, let's assume a user's traffic (say, an email message) is relayed from the user's computer to the recipient's computer. In traditional internets (those that do not use label switching), the method to relay this email is similar to postal mail: a destination address is examined by the relaying entity (for our work, a router; for the postal service, a mail handler). This address determines how the router or mail handler forwards the data packet or the mail envelope to the final recipient.

Label switching is different. Instead of a destination address being used to make the routing decision, a number (a label) is associated with the packet. In the postal service analogy, a label value is placed on the envelope and is thereafter used in place of the postal address to route the mail to the recipient. In computer networks, a label is placed in a packet header and is used in place of an address (an IP address, usually), and the label is used to direct the traffic to its destination.

WHY USE LABEL SWITCHING?

Let's look at the reasons label switching is of such keen interest in the industry. We examine the topics of (a) speed and delay, (b) scalability, (c) simplicity, (d) resource consumption, and (e) route control.

Speed and Delay

Traditional IP-based forwarding is too slow to handle the large traffic loads in the Internet or an internet, a topic explained in Appendix A. Even with enhanced techniques, such as a fast-table lookup for certain datagrams, the load on the router is often more than the router can handle. The result may be lost traffic, lost connections, and overall poor performance in an IP-based network.

Label switching, in contrast to IP forwarding, is proving to be an effective solution to the problem. Label switching is much faster because the label value that is placed in an incoming packet header is used to access the forwarding table at the router; that is, the label is used to index into the table. This lookup requires only one access to the table, in contrast to a traditional routing table access that might require several thousand lookups.

The result of this more efficient operation is that the user's traffic in the packet is sent though the network much more quickly than with the traditional IP forwarding operation, reducing the delay and response time to enact a transaction between users.

Jitter. For computer networks, speed and its nemesis, delay, have another component. It is the variability of the delay of the user traffic, caused by the packet traversing several to many nodes in the network to reach its destination. It is also the accumulation of this variable delay as the packet makes its way from the sender to the receiver. At each node, the destination address in the packet must be examined and compared to a long list of potential destination addresses in the node's (usually a router) routing table.[1]

---

[1]The terms routing table and forwarding table are used synonymously in this book.

As the packet traverses through these nodes, it encounters both delay and variable delay, depending on how long it takes for the table lookup and, of course, on the number of packets that must be processed in a given period. The end result, say, at the receiving node, is jitter, an accumulation of the variable delays encountered at each node between the sender and the receiver.

This situation is onerous to speech packets because it often translates into uneven speech play-out to the person listening to the speech. It may even result in a person's having to wait a few seconds to receive the final words of a sentence as the speech packets make their way through the network.

Once again, the more efficient label switching operation results in the user's traffic being sent through the network much more quickly and with less jitter than with the traditional IP routing operation.

Scalability

Certainly, speed is an important aspect of label switching, and processing the user traffic quickly in an internet is very important. But fast service is not all that label switching provides. It can also provide scalability. Scalability refers to the ability or inability of a system, in this case the Internet, to accommodate a large and growing number of Internet users. Thousands of new users (and supporting nodes, such as routers and servers) are signing on to the Internet each day. Imagine the task of a router if it has to keep track of all these users. Label switching offers solutions to this rapid growth and large networks by allowing a large number of IP addresses to be associated with one or a few labels. This approach reduces further the size of address (actually label) tables and enables a router to support more users.

Simplicity

Another attractive aspect of label switching is that it is basically a forwarding protocol (or set of protocols, as we shall see). It is elegantly simple: forward a packet based on its label. How that label is ascertained is quite another matter; that is, how the control mechanisms are implemented to correlate the label to a specific user's traffic is irrelevant to the actual forwarding of the traffic. These control mechanisms are somewhat complex, but they do not affect the efficiency of the user traffic flow.

Why is this concept important? It means that a variety of methods can be employed to establish a label *binding* (an association) to the user's traffic. But after this binding has been accomplished, then label switching operations to forward the traffic are simple. Label switching operations can be implemented in software, in application-specific integrated circuits (ASICs), or in specialized processors.

Resource Consumption

The control mechanisms to set up the label must not be a burden to the network. They should not consume a lot of resources. If they do, then their benefits are largely negated. Fortunately, label switching networks do not need a lot of the network's resources to execute the control mechanisms to establish label switching paths for users' traffic (if they do consume a lot of resources, they are not designed well). We spend quite a lot of time in this book examining the control mechanisms to support label switching.

Route Control (Control of the Forwarding Path)

With some exceptions, routing in internets is performed with the use of the IP destination address (or in a LAN, with the destination MAC address). Certainly, many products are available that use other information, such as the IP type of service (TOS) field and port numbers, as part of the forwarding decision. But destination-routing (the destination IP address) is the prevalent forwarding method.

Destination-routing is not always an efficient operation. To see why, consider Figure 1–1. Router 1 receives traffic from routers 2 and 3. If the IP destination address in the arriving IP datagram is for an address found at router 6, the routing table at router 1 directs the router to forward traffic to either router 4 or 5. With some exceptions, no other factor is involved.[2]

Label switching permits the routes through an internet to be subject to better control. For example, a labeled packet emanating from router 2 may be destined for an address at router 6; likewise, the same situation could hold for a labeled packet starting at router 3. However,

---

[2]Some vendors who manufacture routers, bridges, etc., have implemented their own proprietary alternatives to destination-based routing. Some products allow the network administrator to load-level traffic across more than one link; others use the TOS field, port numbers, and so on.

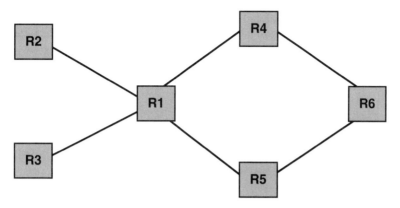

Figure 1–1 Destination-based routing.

the packets' different label values can instruct router 1 to send one la-beled packet to router 4 and a packet with a different label value to router 5.

This concept provides a tool to engineer the nodes and links to ac-commodate traffic more efficiently, as well as to give certain classes of traffic (based on QOS needs) different levels of service. Perhaps the links between routers 1 and 4 and routers between 1 and 5 are DS3 and SONET, respectively. If the user application needs more bandwidth, the user's label can be used to instruct the router to place the traffic on the SONET link instead of the DS3 link. This policy-based approach uses label switching to tailor the network to the needs of the traffic classes, a concept called *traffic engineering* (TE).

Route Control Using IP

As noted, it is possible to use the TOS field and the precedence bits in this field to ameliorate some of the problems associated with destination routing. However, the precedence bits are used in some net-works and not used in others. Although standards define the use (the old RFC 791), a router may leave the bits untouched or may alter them. Likewise, a router may examine them or may ignore them. Notwith-standing these comments, routers can be configured to use the prece-dence bits. Shortly, the use of these bits is described.

For this discussion, it is a good idea to reset these bits for all traffic entering your network from an unknown network. By resetting the precedence values, you guarantee that users who have set these bits to get special treatment (better service) do not receive this service at the expense of your internal network customers.

Policy-Based Routing with IP.   Continuing the discussion on route control, policy-based routing (PBR) is often associated with label switching protocols, such as Frame Relay, ATM, or MPLS. It can also be implemented with IP by using the TOS field as well as port numbers, the IP protocol ID, or the size of the packets.

Using these fields allows the network provider to classify different types of traffic, preferably at the edge of the network; that is, the ingress to the network. Then the core routers can use the precedence bits to decide how to handle the incoming traffic. This "handling" can entail using different queues and different queuing methods.

IP policy-based routing also allows the network manager to execute a form of constraint-based routing, an operation explained in considerable detail in this book. Based on whether the packet meets or does not meet the criteria just discussed, policies can be executed that enable a router to do the following:

- Set the precedence value in the IP header.
- Set the next hop to route the packet.
- Set the output interface for the packet.

We should note that some of the literature in the industry states that IP, without label switching, is not capable of policy-based and constraint-based routing. These claims are not accurate. The problem with the operations described in this section is not that they do not work; indeed they do. But the fact remains that the public Internet consists of many networks and many Internet Service Providers (ISPs), and there is no agreement among these parties on how to use the IP precedence bits.

The same situation potentially holds for label switching. Like IP precedence operations, label switching is only as effective as the agreements among network operators on how it is used.

IP Source Routing. Another way to control the route in IP-based networks is to use the options field in the IP datagram header to support source routing. The sender of the datagram (the source) places a list of IP addresses in the options field, and these addresses are used (one after the other in the list) to specify the route of the datagram. The problem with IP source routing is that (a) it requires the sender to know the route in the first place (which in many networks is not feasible), (b) the list of 32-bit addresses can translate into considerable overhead, and (c) many networks do not support the source routing option.

Nonetheless, IP source routing can be useful in private internets, so it should not be ignored. It comes in two flavors. The first, called *loose source routing*, gives the IP nodes the option of using other intermediate hops to reach the addresses obtained in the source list as long as the datagram traverses the nodes listed. Conversely, *strict source routing* requires that the datagram travel only through the hops whose addresses are indicated in the source list. If the strict source route cannot be followed, the originating host IP is notified with an ICMP error message.

THE ZIP CODE ANALOGY

To understand more about the basic ideas of label switching, let's return to the postal system example. As depicted in Figure 1–2(a), a piece of mail is being forwarded though the postal system from one party to another. Notice that the actual address of the mail recipient is not used in the postal "network" to relay the envelope. Rather, the ZIP code 88888 is used as a label to identify where the mail is to go. After the envelope reaches its destination ZIP area (the end of the "postal path"), then the address (street number, etc.) is used to forward the mail to its intended reader.

This idea holds for label switching. In Figure 1–2(b), an IP datagram (packet) is sent to a label switching router for delivery to a destination IP address. The router appends a label to the packet (something like a ZIP code). Thereafter, the label, not the IP address, is used in the network to forward the traffic. Once the traffic has reached the end of the "label path," the IP address is used to make the final delivery to the end user.

Thus in both networks, the cumbersome addresses are not processed. This common sense approach saves a great deal of time and substantially reduces the overhead of both the postal network and an internet. This reduction of time and overhead is the essence of label switching technology, an indispensable tool in today's internets. In later discussions, we show more detailed examples of IP forwarding and label switching operations and explain further why label switching is so effective.

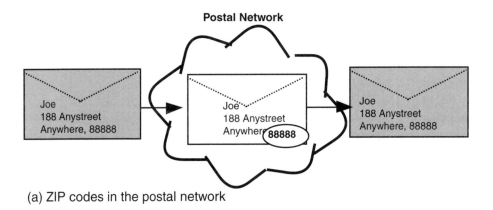

(a) ZIP codes in the postal network

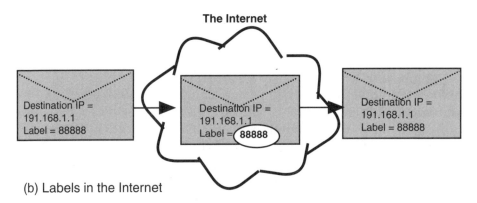

(b) Labels in the Internet

Figure 1–2   ZIP codes and labels.

WHY A LABEL IS NOT AN ADDRESS

A label is not an address. It has no inherent topological significance. Moreover, until the label is correlated with an address, it has no routing significance. Therefore, a requirement still exists for conventional IP address advertising, as shown in Figure 1–3. Part of the job of a label switching network is to correlate the addresses and routes with labels.

The routes are discovered by the IP routing protocols and are based on IP addresses. In this example, the label switching routers are advertising address 191.168.1.1. In most situations, an address prefix is advertised (a prefix is the network and subnetwork part of the 32-bit address), but that need not concern us for this general example.[3] This advertisement reaches the router on the left side of Figure 1–3. The router

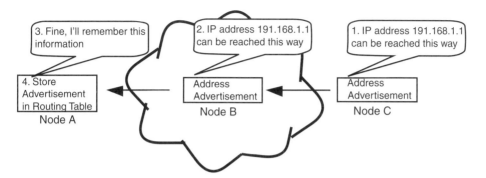

Figure 1–3 Address advertisement.

stores the routing information in its routing table. Thereafter, when the router receives a packet destined for address 191.168.1.1, it consults its routing table to find out how to reach this address.

In a label switching network, an important job is to choose a label value to place onto the packet header for use in the network and to inform the other label switching routers about the association of the label value to the address. How this operation is accomplished is shown in a general way in Figure 1–4. Router (node) A informs router (node) B that address 191.168.1.1 is to be associated with label 88888. This association is called a *bind*.

When router B receives this label/address advertisement, it consults its routing table and looks up the next node that is to receive traffic destined for 191.168.1.1. As we learned in Figure 1–3, that next node is router C. Therefore, router B builds an entry in another table (called by various names: label switching table, label mapping table, cross-connect table, as examples) specifying that an incoming label from node A with a value of 88888 is to be routed onto the outgoing link to node C. This process continues until the packet reaches the final destination.

You may have noticed that I did not show the operations between router B and router C in Figure 1–3. The reason for this exclusion is that there are some additional operations between the LSRs B and C that are explained later.

---

[3]The first part of this book uses complete 32-bit addresses. Later, examples of address prefixes are shown. Keep in mind that part of the value of the use of a label is to correlate one label value to multiple IP addresses. Thus, mapping labels to a prefix and not to an individual address is the sensible way to use label switching. If you are not familiar with address prefixes, take a look at Appendix A.

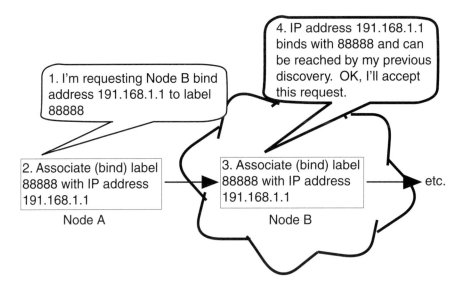

Figure 1–4   Label/address advertisement.

The operation in Figure 1–4 has the label assigned by LSR A after it has discovered the path to the address. Another approach is for the binding to occur at the same time the address is advertised. Consequently, in Figure 1–5, the process of binding begins at node C. The label switching networks can support both approaches, the pros and cons of which are explained in later chapters.

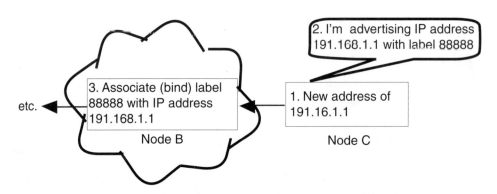

Figure 1–5   Advertising and binding at the same time.

HOW LABEL SWITCHING IS IMPLEMENTED
AND HOW IT CAME ABOUT

Several methods are employed to implement label switching. For this book, we examine those that are deployed and those under consideration, concentrating on the MPLS specifications. As we will see, many of them are similar. Chapter 3 provides a taxonomy of these methods.

The concept of label switching has been around for a number of years, and several firms developed proprietary label switching schemes for their products. These schemes are covered in a companion book to this series, *Advanced Internet Technologies,* and are explained in a general way in Chapter 3. For this book, the cogent aspect of this discussion is to emphasize that the implementation of these proprietary approaches provided a wealth of experience and information about label switching. But the proprietary schemes are not compatible. Therefore, the Internet Engineering Task Force (IETF) set up a working group to establish a standard for the label switching technology. This standard is MPLS. We emphasize MPLS throughout this book with less emphasis on the proprietary label switching protocols, since the proprietary protocols will be replaced by MPLS.

CLARIFICATION OF TERMS

Later chapters explain the major responsibilities of internetworking units, such as label switching routers, but it is important to pause here and clarify some terms. Two protocols are employed by routers to successfully relay the user traffic to its receiver: (a) one protocol (say, protocol 1) relays packets from a source user to a destination user, and (b) the other protocol (say, protocol 2) finds a route for the packets to travel from the source to the destination.

Unfortunately, several terms are used to describe these two types of protocols, and the terms themselves are not models of clarity. Nonetheless, we must deal with them at the onset of our journey through label switching networks; otherwise, many parts of this book will be quite confusing. Figure 1–6 is used to explain these terms.

The older term to describe protocol 1 is *routing,* and the older terms to describe protocol 2 are *route advertising* or *route discovery.* These two terms are still used in the industry in the context just described.

Today, as Figure 1–6 shows, the term *routing* describes protocol 2, and the terms *forwarding* and *switching* describe protocol 1. In keeping with current industry practice, the new terms are used in this book. I use

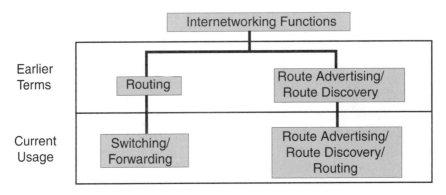

Figure 1–6   Terms and concepts.

route advertising, route discovery, and routing synonymously. However, I continue to use the term *routing table* or *label table* to describe the table of addresses or labels used to forward packets through the network.

To summarize, two protocols are involved in the internetworking process:

- Forwarding/Switching: Using a routing table or a label table to make a forwarding decision.
- Routing: Using route advertisements to acquire the knowledge to create the routing/label table that the forwarding protocol uses. For label switching networks, this advertising may entail the advertising of an address and its associated label.

THE NEED FOR A QOS-BASED INTERNET

The provisioning of adequate resources for an application (such as bandwidth for fast relay through the network) is not a simple process. Because of its complexity, internets in the past treated all applications' traffic alike and delivered the traffic on a best-effort basis, something like the postal service does for regular mail. That is, the traffic was delivered if the network had the resources to support the delivery. However, if the network became congested, the traffic was simply discarded. Some networks have attempted to establish some method of feedback (congestion control) to the user in order to request that the user reduce the infusion of traffic into the network. But as often as not, this technique is ineffective because many traffic flows in data networks are very short, maybe just a few packets in a user-to-user session. So, by the time the user application receives the feedback, it has finished sending traffic. The feedback packets are worthless and have done nothing but create yet more traffic.

The best-effort concept means traffic is discarded randomly; no attempt is made to do any kind of intelligent traffic removal. This approach results in more packets being discarded from applications that require high bandwidth and that place more packets into the network than are discarded from applications with lesser requirements and fewer packets sent into the network. So, the biggest "customers," those needing more bandwidth, are the very ones who are the most penalized! Assuming the customer who is supposedly given a bigger "pipe" to the network is paying more for that pipe, then it is reasonable to further assume that this customer should get a fair return on his or her investment.

It is charitable to say that the best-effort approach is not a very good model. What is needed is a way to manage the QOS in accordance with the customer's requirements and investment.

Label Switching and QOS

In the past few years, it has become increasingly evident that internets need to differentiate between types of traffic and to treat each type differently. We will have more to say shortly about this need, but for this discussion, we need first to define quality of service. The term was first used in the Open Systems Interconnection (OSI) reference model to refer to the ability of a service provider to support a user's application requirements with regard to bandwidth, latency (delay), jitter, and traffic loss. You may notice that these categories[4] are quite similar to the list of reasons for the use of label switching, discussed earlier.

The provision of bandwidth for an application means the network has sufficient capacity to support the application's throughput requirements, measured, say, in packets per second.

The second service category is latency, which describes the time it takes to relay a packet from a sending node to a receiving node. Another term for latency is *round-trip time* (RTT), which is the time it takes to send a packet to a destination node and receive a reply from that node. RTT includes the transmission time in both directions and the processing

---

[4]Other QOS categories, such as security, pricing, and service agreements, are beyond the scope of this book.

time at the destination node. Applications, such as voice and video, have strict latency requirements. If the packet arrives too late, it is not useful and is ignored, resulting in wasted bandwidth and a reduction in the quality of the service to the application.

The third service category, jitter, was discussed earlier. It is the variation of the delay between packets and usually occurs on an output link, where packets are competing for the router's outgoing links' bandwidth. Variable delay is onerous to speech. It complicates the receiver's job of playing out the speech image to the listener.

The last service category is packet loss. Packet loss is quite important in voice and video applications, since the loss may affect the outcome of the decoding process at the receiver and may also be detected by the end user.

The Contribution of Label Switching

You might ask what label switching has to do with QOS. It does not have anything to do with certain aspects of the QOS categories, such as raw bandwidth. However, I stated earlier that label switching can be a valuable tool to combat latency and jitter, two important QOS operations for delay-sensitive traffic, such as video and voice, and for fast Web responses. Since label switching speeds up the relaying of traffic in an internet, it follows that the technology will reduce latency and improve jitter. Indeed, an internet that does not use label switching runs the risk of experiencing unacceptable QOS performance for delay-sensitive traffic.

Of course, label switching unto itself will not solve the delay and variable delay problems that are systemic to data networks. If we are connected to a low-bandwidth network, label switching is not going to give us more bandwidth, but I am stating label switching will ameliorate delay and jitter problems significantly.

LABEL SWITCHING'S LEGACY: X.25 AND VIRTUAL CIRCUITS

For a change of pace here, take a look at a bit of history. The label switching concept began with X.25. In the late 1960s and early 1970s, many data communications networks were created by companies, government agencies, and other organizations. The design and programming of these networks were performed by each organization to fulfill specific business needs. During this time, an organization had no reason to adhere to any common convention for its data communications protocols, since the organization's private network provided services only to itself. Consequently, these networks used specialized protocols that were tailored to satisfy the organization's requirements.

During this period, several companies and telephone administrations in the United States, Canada, and Europe implemented a number of *public data networks* based on packet switching concepts. These systems were conceived to provide a service for data traffic that paralleled the telephone system's service for voice traffic.

But they did not nail up bandwidth as the telephone system did. Indeed, X.25 represented a major change in viewing service to a user: use a best-effort approach but allow the user to request certain levels of service.

The public network vendors were faced with answering a major question: How can the network best provide the interface for a user's terminal or computer to the network? The potential magnitude of the problem was formidable because each terminal or computer vendor had developed its own set of data communications protocols. Indeed, some companies, such as IBM, had developed scores of different protocols within their own product lines.

X.25 came about largely because the originators of these nascent networks recognized that a common network interface protocol was needed, especially from the perspective of the network service providers.

In 1974, the (former) CCITT issued the first draft of X.25 (the "Gray Book"). It was revised in 1976, 1978, 1980, and again in 1984 with the publication of the "Red Book" recommendation. Until 1988, X.25 was revised and republished every four years. In 1988, the ITU-T announced its intention to publish changes to its recommendations (including X.25) as publishing was warranted, rather than in the four-year cycle previously utilized.

The Logical Channel Number: Precursor to the Label

X.25 identifies each packet in the network with a logical channel number (LCN). The LCN is used to distinguish the different users' traffic that is operating on the same physical link. This idea is to mask from the user the fact that the link is being shared by other users, thus the term *virtual circuit* (you think you have the full bandwidth of the link, but you don't). A virtual circuit and its *label,* the logical channel number, are quite similar to the modern label switching network. But I will point out differences as we move into the details of label switching networks.

Frame Relay and ATM: A Rose by Any Other Name Is Still a Rose

The successors to X.25, Frame Relay and ATM, also use the virtual circuit concept. For Frame Relay, the virtual circuit IDs are called data link connection IDs (DLCIs); for ATM, they are called virtual path IDs/virtual channel IDs (VPIs/VCIs). Regardless of their names, they are (a) virtual circuit IDs and (b) label values.

MPLS networks must interwork with these networks, since they are quite prevalent as the principal bearer of services for wide area internets. Fortunately, the MPLS labels correlate rather easily with the ATM and Frame Relay labels, and later chapters explore this subject in considerable detail.

MPLS: STATUS AND CONCEPTS

The Multiprotocol Label Switching (MPLS) Protocol is published by the Internet Engineering Task Force as RFC 3031.

MPLS is a label swapping (mapping) and forwarding technology, but it integrates label swapping with network layer routing. Label swapping, or mapping, means the changing of the label value in the packet header as the packet moves from one node to another; the rationale for this operation is explained in Chapters 2 and 4.

The idea of MPLS is to improve the performance of network layer routing and the scalability of the network layer. An additional goal is to provide greater flexibility in the delivery of routing services (by allowing new routing services to be added without a change to the forwarding paradigm).

MPLS does not make a forwarding decision with each L_3 datagram but uses a concept called the forwarding equivalence class (FEC). An FEC is associated with a class of datagrams; the class depends on a number of factors, at a minimum the destination address and perhaps the type of traffic in the datagram (voice, data, fax, etc.). Based on the FEC, a label is then negotiated between neighbor LSRs from the ingress to the egress of a routing domain. As we showed earlier, the label is then used to relay the traffic through the network.

The initial MPLS efforts of the IETF focus on IPv4 and IPv6. The core technology can be extended to multiple network layer protocols, such as IPX, and SNA. However, there is little interest in expanding MPLS to other network layer protocols, since IP is by far the most pervasive.

The basic idea is not to restrict MPLS to any specific link layer technology, such as ATM or Frame Relay. Most of the efforts so far are directed

to the interworking of MPLS and ATM, but in the future, MPLS will operate directly with IP over the physical layer and not use ATM at all.

In addition, MPLS does not require one specific label distribution
protocol (agreeing on the use of label values between neighbor LSRs). It
assumes there may be more than one, such as the Resource Reservation
Protocol (RSVP), Border Gateway Protocol (BGP), or the Label Distribution Protocol (LDP). Considerable attention is on LDP, since it is being
designed from scratch for MPLS networks. Other protocols, such as BGP
and RSVP, are also good methods for label distribution.

EXAMPLES OF LABEL AND QOS RELATIONSHIPS

We have learned that the label is used for forwarding operations—to determine how to relay the packet to the next node. We also learned that it
can be used to determine the services that will be provided to the packet
during its journey through the network. Thus, the label may be associated with the packet's QOS support. The words "may be" must be emphasized because some MPLS implementations use the label to make QOS
decisions and some do not.

Figure 1–7 shows the way in which two packets are processed at an
LSR and the relationships of QOS and label operations. The packets,
identified with labels 30 and 70, are sent to the switch's interface from
an upstream node (say, another router, not shown in this figure). The labels are then used to access a label switching table. The two table entries
for labels 30 and 70 are shown at the bottom of the figure.

Each table entry contains the label number and the associated
ingress interface number: 30.a and 70.a for these two packets. The
ingress interface is the communications link interface on the router. The
table information is associated with the profile conformance entries and
is used to monitor the flow associated with each of these packets. These
profile examples are usually relevant to the first switch in the QOS domain; that is, at the user-to-network interface (UNI). They determine if
the packet flow is conforming or nonconforming to the service SLA.[5] The
profile for flow 30 is a burst tolerance (BT) of 210 packets per second

---

[5]The SLA is a contract between the network provider and the customer. It stipu-
lates the QOS that the provider agrees to provide, including services not described in this
book, such as security, management reports, and penalties for nonconformance to the
SLA contract.

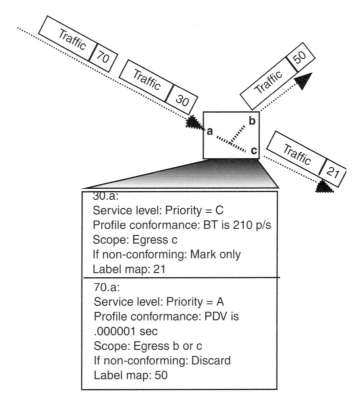

Figure 1–7 Labels and associated QOS operations.

(210 p/s); the profile for flow 70 is a packet delay variation (PDV) of no greater than 1 microsecond.

The service-level entries in the tables also reveal how the packet is to be treated if the traffic for each flow adheres to its SLA. This example uses priorities to differentiate the treatments. Priority = C is for asynchronous traffic, and priority = A identifies synchronous, real-time traffic. The priorities in the example are relative; priority C is not as high as A.

Another entry in the table is the scope of service, shown as "Scope" in the figure. The service for these two packets is scoped to the switch's egress interfaces, the router's output link interfaces. If the services for these packets are end-to-end, perhaps through multiple QOS domains, the ingress and egress must be provisioned at each node that resides in the QOS region. Therefore, we can assume in this example that these egress ports are provisioned with an end-to-end path in mind. Notice that packet 70 is scoped to two egress interfaces, b and c. This approach enables the packet to be routed to the link that is experiencing better performance or to be routed around network failures.

Another entry in the table reflects the operations that are to be performed on the packet if its flow does not conform to the SLA conformance profile. Packet 30 will be marked (tagged) if it is nonconforming. "Mark only" means the packet is not to be discarded (unless the switch is in a precipitous situation). The tag will relegate the service on the packet flow to a lesser quality of service. Packet 70 belongs to a synchronous real-time flow, so if it is nonconforming, it is discarded.

The last entry in the tables is the label that is placed in the packet header for transmittal out of the egress interface to the next node. Label 30 is mapped to label 21, and label 70 is mapped to label 50. The term *map* is also called a *swap*. So, label swapping is the changing of the label value at the LSR. Label swapping is quite important in label switching networks and is explained in Chapters 2, 3, and 4.

DETERMINATION OF THE PHYSICAL PATH THROUGH THE NETWORK: THE LABEL SWITCHED PATH (LSP)

The complete path through a label switched network is called the label switched path (LSP). It is determined in one of two ways. With the first method, traditional routing protocols (such as OSPF or BGP) are used to discover IP addresses. This information, the next node to an address, is correlated with a label, yielding the LSP. With the second method, the LSP can be set up (configured manually) according to the idea of constraint-based routing (CR). This approach may use a routing protocol to assist in setting up the LSP, but the LSP is "constrained" by other factors, such as the need to provide a certain QOS level. Indeed, delay-sensitive traffic is the prime candidate for constrained routing.

The end-to-end LSP is called an LSP tunnel, which is a concatenation of each LSP segment between each node, as shown in Figure 1–8. The characteristics of the LSP tunnel, such as bandwidth allocation, are determined by negotiations between the nodes, but after the negotiation, the ingress node (the beginning of the LSP), shown as node B in Figure 1–8, defines the traffic flow by its choice of the label (L). As the traffic is sent through the tunnel, the idea is not to examine the contents of any other headers but to examine just the label header. Therefore, the remainder of the traffic is "tunneled" through the LSP without examination or alteration. At the end of the LSP tunnel, the egress node (node D

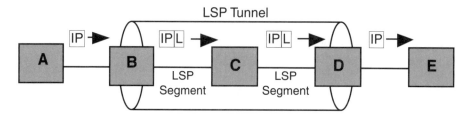

 = Label (L) and IP, including other headers and user traffic

Figure 1–8   LSP tunnels.

in this example) removes the label and passes the IP traffic to an IP node.

As explained in Chapters 7, 8, and 9, LSP tunnels enable the implementation of traffic engineering (TE) policies related to network performance optimization. For example, LSP tunnels can be automatically or manually routed away from network failures, congestion, and bottlenecks. Also, multiple parallel LSP tunnels can be established between two nodes, and traffic between the two nodes can be mapped onto the LSP tunnels according to local policy.

SUMMARY

Traditional IP forwarding is too slow to handle the large traffic loads in the Internet or an internet. Even with enhanced techniques, such as a fast-table lookup for certain datagrams, the load on the router is often more than the router can handle. The result may be lost traffic, lost connections, and overall poor performance in the IP-based network. Label switching, in contrast to IP forwarding, is proving to be an effective solution to the problem. The main attributes of label switching are fast relay of the traffic, scalability, simplicity, and route control.

MPLS represents a vendor-independent specification for label switching. It improves the performance of network layer routing and the scalability of the network layer, and provides greater flexibility in the delivery of routing services.

2

Label Switching Basics

This chapter introduces the basic concepts of label switching. We begin with the examination of control planes, a topic discussed in several parts of this book. We then learn more about the forwarding equivalence class (FEC), introduced in Chapter 1: the information that makes up an FEC and how an edge router associates the FEC with a label and a class of service.

We examine label allocation methods, with examples of local and remote binding, upstream and downstream binding, and control and data binding operations. We are introduced to the concept of a label space, and we see examples of how labels are set up between neighbor routers.

IP AND MPLS CONTROL AND DATA PLANES

In recent Internet RFCs and working drafts, the term control plane has come into use. A control plane is a set of software and/or hardware in a machine, such as a router, and is used to control several vital operations of the network, such as label allocations, route discovery, and error recovery. The job of the control plane is to provide services to the data plane, which is responsible for relaying user traffic through the router.

The term data plane does not mean the traffic is only data; it might be voice or video traffic. The terms user plane and transport planes are also used to describe the data plane.

IP Control and Data Planes

Figure 2–1 shows the relationships of the IP control plane to the IP data plane. For the Internet protocols, examples of control planes are the routing protocols (OSPF, IS-IS, BGP). They enable IP (in the data plane) to forward traffic correctly.

The control messages are exchanged between routers (and other machines that participate in IP routing and forwarding functions) to perform a variety of operations, including:

- Exchanging messages between nodes to establish a routing relationship (including security agreements).
- Exchanging periodic messages (called hellos) to make certain the neighbor nodes are up and running.
- Exchanging route and address advertisement messages to build routing tables used by IP to forward traffic.

In Figure 2–1, the arrow pointing from the control plane to the routing table means that the routes discovered by the routing protocols are stored in the routing table. The bidirectional arrow shown between the routing table and the data plane means IP accesses the routing table to perform its forwarding operations.

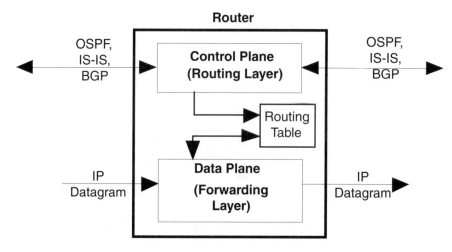

Figure 2–1   IP control and data planes.

MPLS Control and Data Planes

MPLS also operates with control and data planes, as depicted in Figure 2–2. The main job of the control plane is to advertise labels and addresses and to correlate them; that is, to bind (map) labels to addresses. This idea is explained in more detail shortly.

The label switching router (LSR) is a router that has been configured to support MPLS. It uses the label forwarding information base (LFIB) to determine how to process the incoming MPLS packets, such as determining the next node to receive the packet. The LFIB is to MPLS what the routing table is to IP. LFIB is a common name used by Cisco. In later chapters, the LFIB is examined in the context of the MPLS RFC and is given other names.

More than one protocol can operate at the MPLS control plane. As examples, RSVP has been extended to allow the use of this protocol to advertise, distribute, and bind labels to IP addresses. This extension is called RSVP-TE. A protocol known as the Label Distribution Protocol (LDP) is yet another option for executing the MPLS control plane. We will also examine several extensions to OSPF and BGP; they also assist in the operations of the control plane, shown in Figure 2–2 as OSPF-E and BGP-E.

The control messages are exchanged between label switching routers (LSRs) to perform a variety of operations, including:

• Exchanging messages between nodes to establish a relationship (including security agreements). After this operation is complete, the nodes are said to be LSR peers.

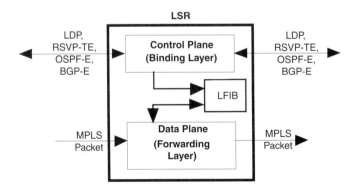

Figure 2–2   MPLS control and data planes.

- Exchanging periodic messages (called hellos) to make sure the neighbor nodes are up and running.
- Exchanging label and address messages to bind addresses to labels and build forwarding tables (notably the LFIB) used by the MPLS data plane to forward traffic.

After MPLS nodes have exchanged labels and IP addresses, they then bind the labels to addresses. Thereafter, the MPLS data plane forwards all traffic by examining the label in the MPLS packet header. The IP address is not examined until the traffic is delivered across the network (or networks) to the receiving user node. The label header is then removed, and the IP address is used by the IP data plane to deliver the traffic to the end user.

What is not shown or explained in this introduction is how the IP and MPLS planes are coordinated with each other. Let's move on, and we will pick up this operation later in the chapter.

THE FORWARDING EQUIVALENCE CLASS

The term FEC is applied to label switching operations. FEC describes an association of discrete packets with a destination address, usually the final recipient of the traffic, such as a host machine. FEC implementations can also associate an FEC value with a destination address and a class of traffic. The class of traffic is typically associated with a destination port number.

Why is FEC used? First, it allows the grouping of packets into classes. From this grouping, the FEC value in a packet can be used to set priorities for the handling of the packets, giving higher priority to certain FECs over others. FECs can be used to support efficient QOS operations. For example, FECs can be associated with high-priority, real-time voice traffic, low-priority newsgroup traffic, and so on.

The matching of the FEC with a packet is achieved by using a label to identify a specific FEC. For different classes of service, different FECs and their associated labels are used. For Internet traffic, the following identifiers are candidate parameters for establishing an FEC. Be aware that in some systems, only the destination IP address is used.

- Source and/or destination IP addresses
- Source and/or destination port numbers

- IP Protocol ID (PID)
- IPv4 Differentiated Services (DS) Codepoint
- IPv6 flow label

Scalability and Granularity

The network administrator can control how big the forwarding tables become by implementing FEC coarse granularity. If only the destination address is used for the FEC (and, of course, address prefixes are used; see Appendix A), the tables can be kept to a manageable size. Yet this coarse granularity does not provide a way to support classes of traffic and QOS operations. On the other hand, a network supporting fine granularity by using port numbers and PIDs will have more traffic classifications, more FECs, more labels, and a larger forwarding table. This network will likely not scale to a large user base.

Fortunately, label switching networks need not be one or the other. A combination of coarse and fine granularity FECs is permissible.

At this stage in the evolution of MPLS in commercial products, the most common practice is for a router to automatically assign a label to every FEC known to the router. In a typical unicast environment, this means that a label is assigned to every IP address prefix in the routing table. Also, when a new route is discovered and entered into the routing table, it is assigned a label. Therefore, IP routing triggers the updates to the label data bases in the router.

Granularity in Terabit Networks

The issue of fine granularity in high-capacity networks may not be a key issue. For example, as the industry evolves to optical-based switching fabrics and as network nodes are connected with more optical fibers that operate at the terabit/s range, the speed of packet processing inside the network is so great that it makes little sense to spend resources (and especially time) in using an FEC to prioritize and queue traffic. Instead, traffic will be forwarded according to destination information; that is, only according to the correlation of labels with destination IP addresses. Inside the network, all traffic is treated the same.

However, the network provider will still want FEC granularity between the network and the user (at the edge of the network) in order to price the network services that offer QOS operations such as priority, throughput, and turnaround time (response time) for the user applications.

Information Used in the Forwarding Decision

Keep in mind that whatever the terms used, the focus of label switching is the *forwarding of a packet to its final destination*. And as we learned, operations may base their forwarding decisions on one or more fields in the incoming packet. These fields are listed here (in more detail than in the list above) and depicted in Figure 2–3. You will note that some of the information explained in this section was not listed above. The reason is that some of these values are used by a router, a switch, or a bridge to make forwarding decisions, but they are usually not used for the FEC.

- Layer 2
  (a) A LAN address (the IEEE MAC address)
  (b) An ATM or Frame Relay virtual circuit ID (VCID)
- Layer 3
  (a) Destination and source IP addresses (or some other layer 3 address, such as IPX, AppleTalk, etc.)
- Layer 4
  (a) Destination and source port numbers
- IP Protocol ID

The reason that port numbers and the IP Protocol ID may be used in the FEC and the forwarding decision process is that these fields (the destination port number and the PID) identify the type of traffic residing in the IP datagram payload. For example, the PID may be coded by the

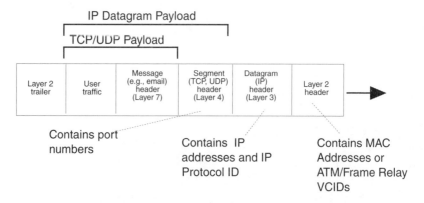

Figure 2–3   Information used in forwarding decisions.

transmitter of the original datagram to indicate that the payload is OSPF traffic. A router can be programmed to treat this traffic differently from the PID indicated the payload was, say, TCP or UDP traffic. If indeed the payload contains TCP or UDP traffic, the port numbers in the TCP or UDP header then indicate what type of TCP or UDP payload resides in the remainder of the packet. For example, the destination port number might be coded to signify that the traffic is voice, email, file transfer, and so on. Thus, these fields become quite important for networks that need to support different QOS services for different kinds of traffic; that is, fine granularity.

The MPLS label is not shown in Figure 2–1, unless it is the ATM or Frame Relay VCID. Another header, called a *shim header,* may exist in the packet and is explained shortly.

LABEL ALLOCATION METHODS

The assignment of the value to a packet varies, depending on the vendor's approach or the standard employed (an Internet RFC or Working Draft). This part of the chapter introduces the concepts of label allocation (binding); in later discussions, we focus on a more detailed examination.

Local and Remote Binding

The term binding refers to an operation at a label switching router (LSR) in which a label is associated with an FEC. Local label allocation (local binding) refers to the operation in which the local router sets up a label relationship with an FEC. The router can set up this relationship as it receives traffic or as it receives control information from a neighbor. A common approach is for a router to simply assign a label to each IP prefix it knows about and then advertise these relationships according to rules (explained in Chapter 4). As shown in Figure 2–4, remote binding is an operation in which a neighbor node assigns a binding to the local node. Typically, this is performed with control messages, such as a label distribution message.

Downstream and Upstream Binding

As depicted in Figure 2–5, downstream label allocation refers to a method by which the label allocation is done by the downstream LSR. The term *downstream* refers to the direction in which a user packet is

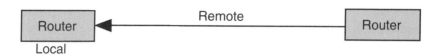

Figure 2–4   Local and remote bindings.

sent. When the upstream router (Ru) sends a packet to the downstream router (Rd), the packet has been identified previously as a member of an FEC and the label (say, label L) is associated with the FEC. Thus, L is Ru's outgoing label, and L is Rd's incoming label.

The term downstream can be a bit confusing. The best way to think of downstream from a router's point of view is to consider a downstream node to be the next hop in the IP routing table of that router to reach an address. So, downstream is a relative term, depending on whether a router is the next hop in a route to a destination.

Consequently, it may be that a router knows about many addresses that are not the next hop for a particular IP datagram. If this is the case, the router would not have to assign a label to these addresses with its upstream neighbors because labels are only used to forward traffic to a next node in the downstream direction.

CONTROL OPERATIONS FOLLOWED BY DATA OPERATIONS

The most common approach in MPLS networks is to use the control plane to set up the bindings and the label switched path. This operation can occur by using a label distribution protocol, or by using craft com-

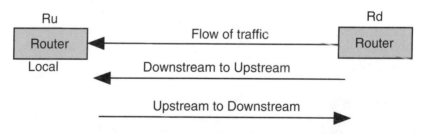

Figure 2–5   Upstream and downstream bindings.

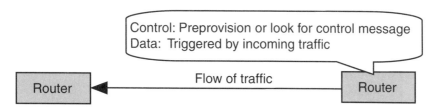

Figure 2–6   Control vs. data-flow-driven bindings.

mands (not recommended because of the laborious nature of manually crafting a network). Many earlier label switching networks support the idea of establishing a binding on-the-fly: as a new address is discovered at a node, the node executes its local binding operations, then its label distribution.

Figure 2–6 reinforces the idea of (a) first invoking the control plane to set up the label switched path and (b) then using the data plane for the ongoing forwarding operations.

LABEL SPACE AND LABEL ASSIGNMENTS

Labels can be assigned between LSRs by one of two methods. In explaining this idea, we use the term *label space* to refer to the way in which the label is associated with an LSR. Figure 2–7 illustrates these ideas.

The first method is a *per-interface label space*. Labels are associated with a specific interface on an LSR, such as a DS3 or SONET interface. Common implementations of this method are ATM and Frame Relay networks, where virtual circuit ID labels are associated with an interface. This approach is used when two peers are directly connected over an interface, and the label is used only to identify traffic sent on one interface. If the LSR uses an interface value to keep track of the labels on each interface, a label value can be reused at each interface. In a sense, this interface identifier becomes an *internal* label in the LSR for the *external* label sent between the LSRs.

The second method is a *per-platform label space*. Here, incoming labels are shared across all interfaces attached to the node. This means the node (such as a host or an LSR) must allocate the label space across all

Per Interface Label Space

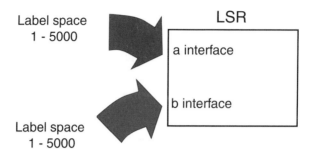

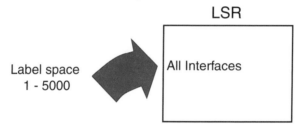

Figure 2–7   Label space and label assignments.

interfaces. The choice for these methods is implementation-specific, although the per-interface label space allocation method is more common as of this writing.

EXECUTION OF THE CONTROL PLANES IN A LABEL SWITCHING DOMAIN

Label switching networks are organized into label switching domains. The domain is made up of one or more networks that belong to an organization such as an ISP or to a private company. Within the domain are label switching routers that are controlled by the domain's administration and that are configured to meet the needs of the domain. Domains communicate with each other through internetworking agreements. For example, America Online may have an internetworking agreement with

AT&T so that these ISPs can pass traffic back and forth between their respective customers.

Figure 2–8 shows one label switching domain and some associated nodes. The user nodes are identified with IP addresses of 172.16.1.1 (the upstream node) and 172.16.2.1 (the downstream node). These nodes likely consist of communications equipment, such as routers. Nodes A and C are called edge LSRs because they sit at the edge (the boundary) of the label switching domain.

From the context of traffic flowing from node 172.16.1.1 to node 172.16.2.1, LSR A is the ingress node to the domain, and LSR C is the egress node. LSR B is known by a variety of names; the most common are (a) transit LSR, (b) interior LSR, or (c) core LSR.

The small letters (a, b, d, a, etc.) placed in the boxes representing the routers identify the physical interfaces in the router. For example, node A has two interfaces, b and d, as shown in Figure 2–8. Interface b could be a DS3 link connecting node 172.16.1.1 to node A, and interface d could be an OC-12 link connecting nodes A and B.

Figure 2–8 shows the major events that take place for all nodes to learn about 172.16.2.1. An IP routing protocol, such as OSPF, performs

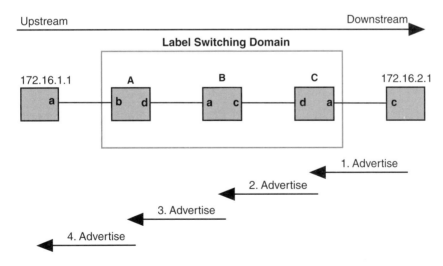

Note: Packet advertisements are OSPF, IS-IS, BGP traffic.

Figure 2–8   Discovering the address and its location.

these services. As a sidenote, it is likely that address prefixes (an aggregation of addresses) are being advertised, but that need not concern us here.

As a result of the completion of the events shown in Figure 2–8, all nodes know about node 172.16.2.1 and about the way to reach the node. They have inserted this information into their routing tables, and in most implementations today, a label is assigned to this address at each of the nodes. The information is kept in a database called the label information base (LIB).

Scenarios for Label Assignments

As noted, after address discovery is complete, a router assigns a label to an address and stores this information in the LIB. Then, label-to-IP address assignments are made to a router's neighbors (the next hops in the routing table). How this operation occurs varies, depending on the routing domain administrator's preference for upstream, downstream, data-driven, or control-driven implementations, as well as on some specific MPLS rules, as discussed in later chapters.

For now, let's look at two scenarios. The first is shown in Figure 2–9. This operation is called solicited label distribution (or downstream-on-demand label distribution), because a label distribution protocol request message is sent before the actual binding message is sent. For this example, the upstream nodes have made this request to the downstream nodes (that is, to the next hop nodes) for address 172.16.2.1.

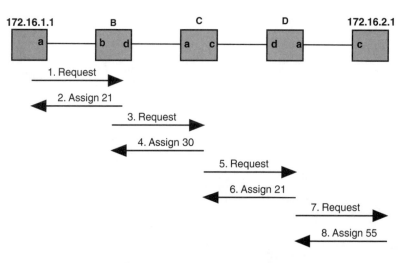

Figure 2–9   Solicited binding.

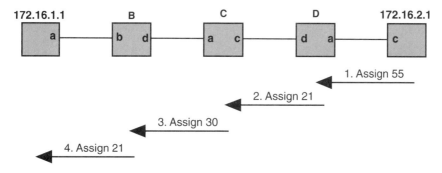

Figure 2–10   Unsolicited binding.

Figure 2–10 shows an example of unsolicited downstream binding. Here, the downstream node on the right side of the figure sends the binding message to its upstream node, which sends the advertisement to the next node in the domain and so on. The upstream nodes have not requested bindings but they receive them anyway.

Unsolicited binding can occur in both an upstream and a downstream direction. Referring to Figure 2–10, node C could send binding

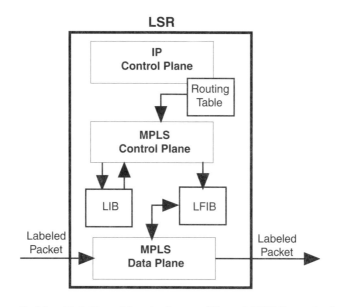

Figure 2–11   Relationships between IP and MPLS control planes.

messages to both nodes B and D. Chapter 4 explains the pros and cons of solicited and unsolicited binding.

Relationships Between IP and MPLS Control Planes

Figure 2–11 shows the relationships between the IP and MPLS control planes. The routing table is used to help create the LFIB. Each time a new IP address prefix is added to the routing table, the router's operating system allocates a new label to it and places this information in the LIB. Using information from the IP routing table and the MPLS LIB, the LFIB is updated, and then used by the MPLS data plane to forward label packets through the node to the next hop on the label switched path.

EXAMPLES OF FEC AND LABEL CORRELATIONS:
THE LABEL SWITCHING TUNNEL

To extend this example, let us refer to the example in Chapter 1 (Figure 1–2). In Figure 2–12, the packet is sent to the edge router; this router examines the FEC-related fields in the headers. It decides to assign a label to this packet as well as to treat the packet in a certain way, such as forwarding the packet to an output queue. The packet is encapsulated into an outer packet, and the header of the outer packet has label number 88888 placed in it.

This idea is called a *label switching tunnel,* which means the inner packet is not examined by the internal LSRs within the network. Their

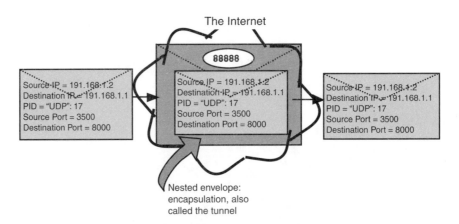

Figure 2–12   Label switching tunnel.

only concern is the processing of the outer packet header's label and handling the packet accordingly. At the egress node, the packet is decapsulated (detunnelled), and the destination IP address, along with the other identifiers, is used to determine how the packet is treated at the receiving node.

ALTERNATIVES FOR CARRYING THE LABEL

To explain basic label switching concepts, I have been somewhat generic in the previous explanations in depicting the tunnel and the contents of the tunneled packet. Figure 2–13 shows a more specific example of a packet that contains a label. The label can reside in one or two headers in the packet. It can reside in the header of the layer 2 bearer services protocol, such as ATM or Frame Relay, it can reside in a special shim header that follows the layer 2 header and precedes the layer 3 (IP) header, or it can reside in both headers, a topic discussed in later chapters.

The choice of how the label is represented in the packet depends on several factors. The principal factor is whether the packet is transported through ATM or Frame Relay. If this transport occurs, the label can reside in the virtual circuit fields in the ATM or Frame Relay headers. If the packet is not transported through these networks, a separate header is used to contain the label.

It is also possible to carry ATM or Frame Relay data units through MPLS nodes; that is, the positions of the shim header and the ATM or Frame Relay headers in Figure 2–13 are reversed. This form of encapsulation is attractive because it provides a graceful way to migrate to MPLS backbone networks that support attached ATM and Frame Relay networks. These operations are explained in Chapter 6.

LABEL SWAPPING

The label (with rare exceptions) does not retain the same value as the packet is transported through the label switching domain. Typically, each LSR accepts the incoming packet and changes the value of the label

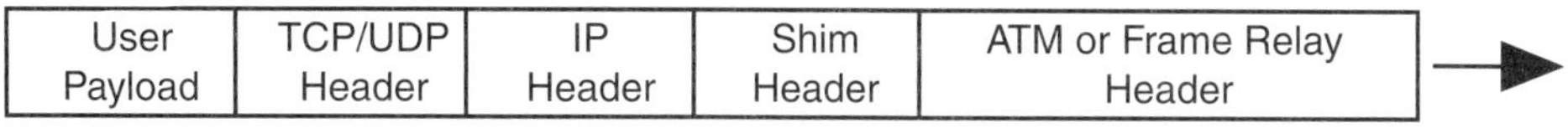

User Payload	TCP/UDP Header	IP Header	Shim Header	ATM or Frame Relay Header	➤

Figure 2–13   Alternatives for carrying the label.

before it sends the packet to the next node in the routing path. The operation is called label swapping (sometimes label mapping).

Figure 2–14 shows some of the entries in the LS tables for one label switching path (LSP) between users 172.16.1.1 and 172.16.2.1. For this discussion, the path is identified as follows:

- Label 21 identifies the LSP between user 172.16.1.1 and switch A.
  a is the output interface at 172.16.1.1
  b is the input interface at switch A
- Label 30 identifies the LSP between switch A and switch B.
  d is the output interface at switch A
  a is the input interface at switch B
- Label 21 identifies the LSP between switch B and switch C.
  c is the output interface at switch B
  d is the input interface at switch C
- Label 55 identifies the LSP between switch C and user 172.16.2.1.
  a is the output interface at switch C
  c is the input interface at 172.16.2.1

Several observations are noteworthy about this figure. First, there must be some means to associate the labels with the FEC, and the association must be made at each machine that participates in the end-to-end LSP.

Second, in this example, the label is correlated with the sender's outgoing interface and the receiver's incoming interface, a point made earlier. Since the labels are so associated, they can be reused at each interface on the switches or user machines. In a sense, the interface numbers in the switch act as *internal* labels for the connection.

Third, the selection of the labels is a matter between the user and its adjacent switch or between adjacent switches. Consequently, there is no requirement to keep the labels unambiguous across interfaces and through the network. For example, label 21 is used twice, between

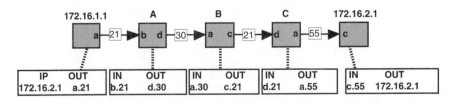

Figure 2–14   LS table entries for an LSP between two users.

172.16.1.1 and switch A, and then between switches B and C. Trying to manage universal label values across multiple nodes and different networks would not be a very pleasant task.

Fourth, the example shows the label bindings (the association of the labels between nodes) in one direction only. It is a straightforward task to use the LS table in a bidirectional manner. For example, if the traffic were flowing from switch C to switch B, the LS table would appear as follows:

- Label 21 identifies the path between switch C and switch B.
  d: is the output interface at switch C
  c: is the input interface at switch B

However, many label switching implementations (MPLS is an example) do not allow a label operation to be bidirectional. This means a two-way connection must have a set of bindings for each direction of the connection.

SUMMARY

This chapter introduced the basic concepts of label switching and further explained the forwarding equivalence class (FEC), introduced in Chapter 1. The chapter then described the information that makes up an FEC and defined how an edge router associates the FEC with a label and a class of service.

We examined label allocation methods and saw examples of local and remote binding, upstream and downstream binding, and control and data binding operations. Later sections of the chapter introduced the concept of a label space and provided examples of how the labels are set up between neighbor routers.

3

Switching and Forwarding Operations

One of the most confusing aspects of switching, routing, and forwarding technologies is discerning exactly what these terms mean. Vendors, standards groups, and service providers often attach different meanings to these terms. We have already dealt with the differences between routing and forwarding (Chapter 1, see Figure 1–6). In this chapter, we clarify the concepts of switching and forwarding by providing a taxonomy of the subject.

Some of the examples in this chapter are proprietary, and some will fade away as MPLS becomes more prevalent. Also, I include quite a lot of material on Cisco's tag switching protocol for two reasons. First, it provides a good example of an actual label switching implementation, and second, it forms the basis for many MPLS operations.

A TAXONOMY OF SWITCHING AND FORWARDING NETWORKS

Initially, the term *routing* referred to making relaying decisions that were performed in a machine typically based with software programs and routing tables stored in conventional RAM. In contrast, *switching* referred to relaying decisions with the support functions existing principally in hardware with specialized processors.

Furthermore, routing traditionally referred to using a destination layer 3 address (for example, an IP address) to make the relaying decisions, whereas switching traditionally referred to using a layer 2 address to perform the relaying operations. In many instances, the layer 2 address was (and still is) a 48-bit IEEE Media Access Control (MAC) address used in local area networks. For layer 3 operations, the address traditionally has been the IP address.

However, in the past few years a number of technologies have emerged that use these techniques or combinations of these techniques and append different names to them. The most common names currently in the industry are described in this chapter. Be aware that many of these techniques are quite similar to one another, and some of them have overlapping functions. As stated, these overlaps make for a confusing mix of techniques.

Figure 3–1 should help you during this discussion. We examine each of the entries in Figure 3–1, starting at the top-left part of the taxonomy and working our way across and down.

Keep in mind that whatever the terms used, the focus of the taxonomy is the *sending of a packet to its final destination.* In addition, the operations described in this chapter base their forwarding decisions on one or more fields in the incoming packet. These fields are described in Chapter 2; see Figures 2–3 and 2–13.

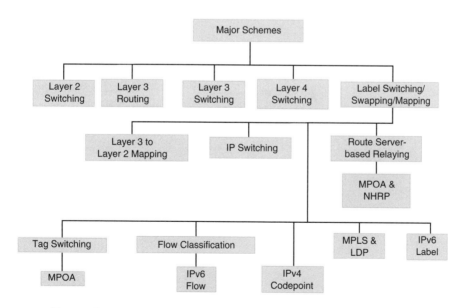

Figure 3–1   A taxonomy for switching and forwarding protocols.

LAYER 2 SWITCHING

Figure 3–2 accompanies the discussions on layer 2 switching and layer 3 routing (forwarding), the subjects of the next two sections in this chapter.

A LAN *bridge* operates at layer 2, the data link layer (always at the MAC sublayer and sometimes at the logical link connection (LLC) sublayer). Typically, the LAN bridge uses 48-bit MAC addresses to perform its relaying functions. The term *layer 2 switching* is often used to describe a LAN bridge.

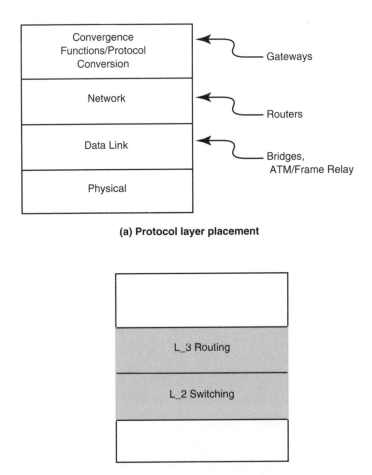

(a) Protocol layer placement

(b) Traditional routing/switching

Figure 3–2   Bridges, routers, gateways, and others.

However, this term is also used to describe an ATM or Frame Relay switch. Since ATM and Frame Relay operate at layer 2, they fit into the category of L_2 switching protocols. Strictly speaking, ATM and Frame Relay should be considered as a combination of L_3 and L_2 switching technologies because both were derived from X.25, which uses a layer 3 header for its principal operations. But most people in the industry use the term L_2 switching for ATM and Frame Relay, so I will defer to this practice.

If ATM or Frame Relay is employed, its virtual circuit IDs (VCIDs) make the forwarding decision. The VCIDs are really labels, although they are managed differently from MPLS labels, a subject discussed in Chapter 6.

LAYER 3 ROUTING (ACTUALLY, FORWARDING)

Layer 3 routing uses a conventional router and forwards the traffic according to a 32-bit IP destination address that resides in the IP header. IP is classified as a layer 3 protocol, thus the term layer 3 routing.

However, it is important that the term routing be clarified once again. In Chapter 1 (see Figure 1–6), we emphasized that routing is now associated with route advertisements and route discovery. Consequently, if you read a recent RFC that explains a routing protocol, the reference is to protocols such as OSPF and BGP, and not to IP.

Problem with IP Forwarding Operations

Traditional IP forwarding operations are fraught with overhead. When the destination address in the IP datagram header is examined by a router, it must match this address against a routing table to determine the next hop to the destination. This operation may require the search of a very large routing table; at major peering points in the Internet, the table is about 50,000 entries. Each incoming packet must be processed against this table. In addition, the use of subnet masks requires the destination IP address in the incoming packet to be matched against the mask (prefix) in the routing table. The longest match rule requires that the route chosen be based on a mask that yields the most matching bits, a topic beyond the scope of this book but explained in a companion book to this series, *IP Routing Protocols*. The result is that conventional IP forwarding simply does not work in large internets. It takes too long to process a packet.

LAYER 3 SWITCHING

Layer 3 switching technology is a newer method of packet forwarding. The distinguishing attribute of layer 3 switching is that the relay functions are performed in hardware through the use of application-specific integrated circuits (ASICs) or specially designed hardware. The relay functions differ from some of the other implementations just discussed in that they do not perform any label mapping nor do they necessarily rely on ATM/Frame Relay-based switching fabrics. The IP address is used without regard to any tag or label.

Cache-Assisted Switching

Some layer 3 switching systems use cache-assisted switching. A cache is built for datagrams containing addresses for specific networks that receive a lot of traffic. With this approach, not every datagram is subject to the conventional reliance on a central table for the route lookup. The most frequently accessed routes are stored in the high-speed cache.

Distributed Switching

With distributed switching, a separate processor is placed on each interface module. The routing table is calculated by a central processor, but the processor does not become involved with the forwarding decisions for each datagram. Instead, the forwarding tables are downline-loaded to the interface processors. In turn, these processors make the forwarding decisions.

Example of Layer 3 Switching

An example of a layer 3 switch is a multigigabit router (MGR) built by BBN Technologies [Part 98].

Figure 3–3 shows the overall architecture for the MGR. We will assume the packets enter the router from the left and exit to the right. The router contains multiple line cards, which support one or more interfaces, and forwarding engine cards, all connected to a switch. The arriving packet has its header removed and passed through the switch to the forwarding engine card, and the other part of the packet is stored on the incoming line card. The forwarding engine examines the header, determines the routing for the packet, updates the header, and forwards it

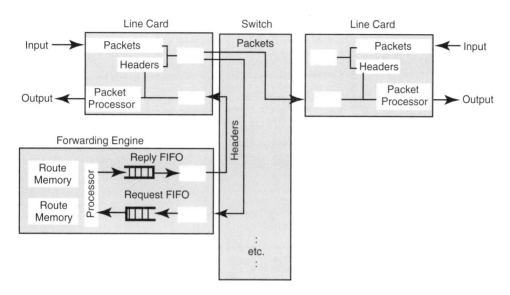

Figure 3–3   Layer 3 switch.

back to the incoming line card, with associated forwarding information. Then, this line card appends the revised header to the other part of the packet and sends the reconstituted packet to the appropriate outgoing line card. The MGR has a total capacity of 50 Gb/s. Its packets per second rate (PPS) is cited at 32 million PPS.

The MGR uses several approaches that are quite different from conventional routers.

First, the router uses distributed routing tables, introduced earlier. Each forwarding engine has a complete set of tables instead of a limited subset of addresses, as found in some routers. This approach avoids the time delay and possible contention in using one central table. Moreover, these tables do not contain all the entries found in conventional tables; they contain only next-hop information.

Second, the switch is not a shared bus, but rather a point-to-point switch containing 15 ports. The switch is an input-queued fabric. Each input maintains a FIFO and uses a protocol to bid for the output. This approach avoids the head-of-line blocking, and the designers state that they have achieved 100 percent throughput.

Third, the forwarding engines are on separate line cards. This approach gives more "real estate" for both functions and allows the designers more flexibility in allocating how many interfaces will share a forwarding engine. Also, it is possible to dedicate a single forwarding engine to a single virtual network, which can simplify configuration and maintenance operations.

Fourth, the line cards are able to accept different L_2 protocol data units, but they must be able to translate them into a common internal L_2 format for processing inside the MGR.

Fifth, the router supports QOS operations. The approach is for the forwarding engine to classify the packet and assign the packet to a flow. This information is passed to the output line card, which schedules the packet transmission with a special QOS processor.

LAYER 4 SWITCHING

Layer 4 switching is a relatively new term. It refers to an operation that examines the Internet port numbers as part of a forwarding decision. The destination port number is certainly used; the source port number may be used. The port numbers are used in conjunction with the source (maybe) and destination (most likely) IP address to make a forwarding decision. The Protocol ID (PID) field in the IP header may also be used. Therefore, layer 4 switching is really not just layer 4 switching. The forwarding operation uses other information as well, an idea explained in Chapter 2 (see Figure 2–3).

LABEL SWITCHING/SWAPPING/MAPPING

The remainder of the taxonomy consists of various renditions of label switching. I title this part of the taxonomy "Label Switching/Swapping/Mapping" to cover the terms used for the operation. Let's take a look at each of these procedures. As you proceed through this book, you will see that MPLS includes many of these "individual" procedures, but not all of them.

LAYER 3 TO LAYER 2 MAPPING

This approach is similar to flow classification and IP switching, with the layer 3 address being mapped to a label or virtual circuit ID. While this example (see Figure 3–4) shows only the mapping of the layer 3 address, the operation can also use Internet port numbers and the PID to derive an FEC for the mapping procedure.

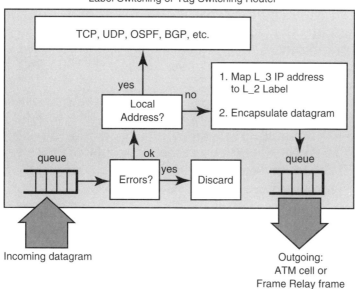

Figure 3–4   Layer 3 to Layer 2 mapping at the ingress LSR.

The address (or FEC) mapping can be to the ATM virtual circuit ID (a virtual path/virtual channel ID (VPI/VCI) or a Frame Relay virtual circuit ID (a data link connection ID (DLCI), or, for that matter, an MPLS label or a Cisco tag. The mapping typically occurs by a router or switch that sits at the edge of the network. Implementations for this method are fairly widespread, including Cisco's tag switching, IBM's ARIS, Cascade's IP Navigator, and Cabletron's Secure Fast Virtual Networking.

At the Ingress LSR

Figure 3–4 shows how an ingress (edge) tag switching router (TSR) or a label switching router (LSR) processes an incoming IP datagram.[1] The incoming packet is stored in a queue to await processing. Once processing begins, the options field in the IP header is processed to determine if any options are in the header (the support for this operation

---

[1]TSR and LSR are different terms that describe the same kinds of operations. Some vendors use the term TSR and others use LSR. I use LSR in this text, unless the specific explanations warrant the use of TSR.

varies; most routers can be configured to examine the type of service (TOS) bits). The datagram header is checked (with a checksum field) for any modifications that may have occurred during its journey to this IP node. The destination IP address is examined. If the IP address is local, the IP PID field in the header is used to pass the data field to the next module, such as TCP, UDP, and ICMP.

If it is determined that the datagram is to be transported through an ATM or a Frame Relay network, the L_3 IP address in the destination field of the IP datagram is correlated to a tag or label that is stored in a table in the LSR. The datagram is then encapsulated into an ATM cell or a Frame Relay frame, with an encapsulation header attached to the datagram.

At an Intermediate (Interior) LSR

The traffic is sent to the outgoing interface for transport to the next node, where the ATM or Frame Relay VCID or a label (and not the IP address) is examined to determine the actions to take on the data unit. This operation is shown in Figure 3–5. The label is examined to determine if it

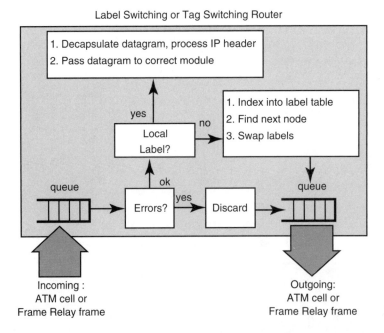

Figure 3–5   Processing at an intermediate or egress LSR.

is local or if it is a binding to the next node. If it is local, the packet is decapsulated and the IP header is used to process the traffic further. If the label indicates the packet is to be relayed to another node, the label is used to index into a label table to find how the packet is to be treated, including its priority, the next node (an egress interface), and the new label that is to replace the old label.

At the Egress LSR

Eventually, the packet will arrive at the final LSR. How the LSR determines that the protocol data unit is at the final node varies, but there must be some method to determine that the label is a local label that "belongs" to that local LSR. Through a process of local binding, which is an operation performed before the transmission of the user data occurs, the local LSR is able to access a table that identifies its labels on each incoming interface.

Therefore, when a cell or frame arrives, the LSR can quickly determine if the label is local; that is, if the traffic is to be terminated at this node and not passed through to the next node. This operation is also shown in Figure 3–5.

The process is straightforward. The ATM cell or Frame Relay header is processed, then removed. The encapsulation header is processed to determine the nature of the user's packet (for example, an IP datagram or an SNA message). Based on the values in the encapsulation header, the packet is passed to the proper module in the LSR or passed to a local machine (such as a router, server, or host) for further processing.

MPLS's Relationship to These Operations

The overall concepts of layer 3 to layer 2 mapping are found in other techniques, such as tag switching, Multiprotocol Over ATM (MPOA), and Multiprotocol Label Switching (MPLS). All of these systems perform some type of L_3 to L_2 mapping—they simply go about it in different ways.

IP SWITCHING

Several methods are employed to implement IP switching; the term was coined by Ipsilon. For this discussion, we examine the use of a high-speed ATM switch, colocated with IP, and the approach used by Ipsilon (purchased by Nokia) and called IP switching.[2] Once again, it will be evident that many of the concepts explained here are present in other procedures in the taxonomy.

First, this approach embodies the concept of a flow, which is a sequence of datagrams from one source machine or application to one or more machines or applications. For a long flow (many datagrams flowing between these entities), a router can cache information about the flow and circumvent the traditional IP routing mechanisms (subnet masking, search on longest subnet mask, and so on) by storing the routing information in cache, thus achieving high throughput.

Generally, the approach is to divert long flows, real-time traffic, or traffic with QOS requirements to an ATM connection and use an individual ATM VPI/VCI for switching the traffic. For transaction-based traffic (database queries, such as a name server operation), the traffic is placed on a preassigned ATM virtual circuit.

One of the difficulties of managing an ATM virtual circuit (VC) is that its state (up, down, inactive, active, etc.) must be maintained end-to-end, with all switches on the VC keeping state information about the VC. If a failure occurs on any part of the VC, extensive signaling is required to clear the VC and take remedial action, such as setting up the VC again. However, if the signaling software is removed and replaced with software that keeps state information only locally, the operations are much more efficient and correlate more closely to the IP's connectionless nature. The technique described here does not use end-to-end VCs; all are local to the two neighbor switches.

Architecture of the IP Switch

Figure 3–6 shows a simple view of the IP switch architecture. The switch contains the ATM switching fabric and IP switch controller, the General Switch Management Protocol (GSMP, see RFC 1987) and Ipsilon Flow Management Protocol (FMP, see RFC 1954).[3] The GSMP gives the IP switch controller access to the ATM switching fabric. The FMP associates IP flows with ATM VCs. The IP switch controller runs conventional

---

[2]For a more detailed description of this approach, see [NEWM98].

[3]Recent documents exclude Ipsilon's name from this protocol and use the initials FMP. I will follow this convention hereafter.

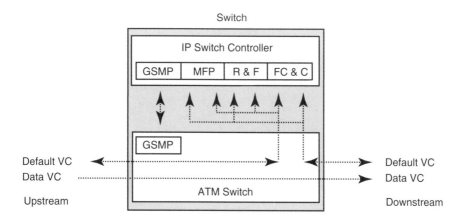

Figure 3–6 IP switch architecture.

IP routing and forwarding operations, in addition to GSMP, FMP, and flow classification and control operations.

IP switching classifies a flow in two ways. The *host-pair flow* uses the source and destination IP addresses and time to live (TTL) to identify the flow. The *port-pair flow* identifies traffic flowing between the same source and destination ports, the same source and destination IP addresses, the same type of service (TOS), the same protocol ID (PID), and the same TTL. At first glance, one might question why all these fields are used. Why not just use the IP addresses? The answer is that identifying ports, protocols, and so on allows a switching decision to be made on the type of traffic—for example, a well-known port such as file transfer or the Domain Name System (DNS). Thus, the file transfer traffic could be switched, and the DNS traffic could be routed. It makes little sense to build a VC for a one-time DNS query.

The host-pair flow is also called a type 2 flow. The *port-pair flow* is called a type 1 flow. So, the definition of a flow in IP switching depends on the type of flow but includes a set of packets whose header fields are identical.

To get things started at system boot, a default VC is established between the switch (switch B) and all its neighbors (say, switches A and C). These VCs forward IP datagrams on a hop-by-hop basis between switches. This VC is shown as the default VC in Figure 3–7 and is identified with a well-known ATM VPI/VCI value. Since this default VC is provisioned as an ATM PVC (permanent virtual circuit), there is no requirement to use the ATM signaling protocol Q.2931. The default VC is

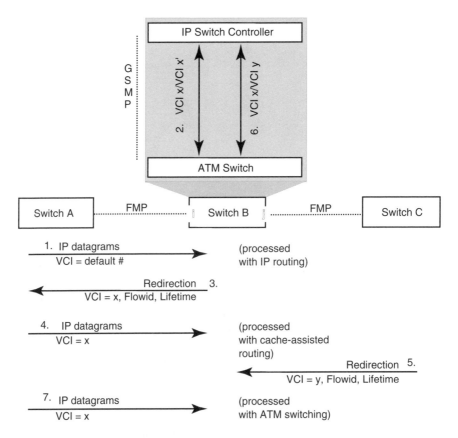

Figure 3–7   Flow redirection.

also used to transfer packets that do not have a label associated with them. These packets are routed by the switch controller routing and forwarding module.

The FMP Operations.   In Figure 3–7, we assume that datagrams arrive at switch A from switch B on the default VC at port i (event 1 in the figure). The IP switch controller (executing AAL5) reassembles the datagram and forwards it with conventional IP routing operations. We further assume that the IP switch controller decides that the datagrams belong to a new flow. In event 2, the controller selects an unused VCI label (say, x′) to operate between the ATM switch and the controller at port c (this port is some type of association between the controller and the switch and is not shown here). The controller selects an unused VCI (say, x) from a table associated with the input port (port i). Next, a switch driver is instructed to map VCI x on port i to VCI x' on port c. The value x' will be used as an index to the cache that has stored the relay information for this VC. The entry is created with GSMP.

In event 3, the switch controller sends an FMP message to the upstream node that has been transmitting this flow. This message contains the label VC = x, a flow label, and a lifetime value. The flow label contains the header fields that set up the flow to begin with. The lifetime states how long the flow is valid.

For a brief period, the cells will arrive at port i with VCI = x, shown as event 4. These cells are mapped to VCI x', passed on port c to the controller, and forwarded to the next node. However, conventional routing table lookups are not performed because X' is used as an index into the cache to obtain the forwarding information.

This intermediate step is needed until the next step occurs. Switch C has been receiving these datagrams and (in event 5) sends a redirection FMP message to its upstream neighbor (switch B), instructing it to redirect this flow to another VCI; in this example, VCI = y. This redirect is on port j. Upon receiving this message, the controller instructs the driver to map x.i to y.j, shown as event 6. Thereafter, the traffic on this flow is no longer processed by the IP switch controller and conventional IP routing. Rather, the traffic is processed directly through the ATM switch to the output port, as depicted in event 7 in the figure.

ROUTE-SERVER-BASED RELAYING

This approach is quite similar to the layer 3 to layer 2 address mapping. The main difference is that a designated machine performs route calculations in contrast to the layer 3 to layer 2 operation in which the translation is performed in the same machine. Examples of this operation are MPOA and the Next Hop Resolution Protocol (NHRP). The route-server-based operations still perform layer 3 to layer 2 address translations—it is simply done in a different machine (a server).

Route-Server-Based Operations

In Figure 3–8, a server residing in Network 1 is responsible for the support of the ability of the client's users to reach destination nodes. The sources and sinks of the traffic are IP nodes, using conventional IP addresses, and the client might be a local router.

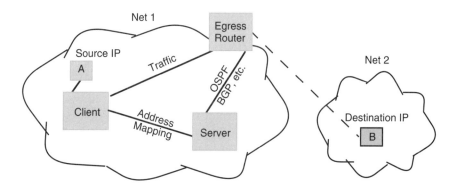

Figure 3–8   Route-server-based relaying.

The client is not responsible for route discovery outside its routing domain to, say, network 2. This task is assumed by an egress router, running an external gateway protocol, such as the Border Gateway Protocol (BGP). The addresses discovered by the egress router are conveyed to the server.

Later, when user A sends traffic to user B, user A's datagram is intercepted by the client. The client's responsibility is to send this traffic to the most efficient egress point in the network. It may have this information stored locally. If not, it sends a request to the server.

The server knows about the destination IP address, courtesy of the egress router's keeping the server informed through the Open Shortest Path First, (OSPF), BGP, or some other routing protocol. The server will respond to the client's request by sending the client a message containing the address of the egress router. The client then forwards this traffic for user B to the egress router, typically over an established ATM connection. The server could also send a label that is to be used for this particular flow.

Multiprotocol Over ATM (MPOA) and Next Hop Resolution Protocol (NHRP)

The principal objective of MPOA is to support the transfer of inter-subnet unicast traffic. MPOA allows the inter-subnetwork traffic based on layer 3 protocol communications to occur over ATM virtual channel connections (VCCs) without requiring routers in the data path. The goal of MPOA is to combine bridging and routing with ATM in a situation where diverse protocols and network topologies exist.

The job of MPOA is to provide this operation to allow the overlaying of layer 3 protocols (also called internetwork layer protocols) on ATM.

MPOA is designed to use both routing and bridging information to locate the optimal route through the ATM backbone.

MPOA supports the concept of virtual routing, which is the separation of internetwork layer route calculation and forwarding. The idea behind virtual routing is to enhance the manageability of internetworking by decreasing the number of devices that are configured to perform route calculation. In so doing, virtual routing increases scalability by reducing the number of devices that participate in the internetwork layer route calculations.

MPOA is responsible for five major operations:

1. *Configuration.* This operation obtains configuration information from the emulated LAN configuration servers (ELAN LECs). These nodes are not important to this discussion; they contain timer information, requirements for the size of packets, etc.

2. *Discovery.* MPOA clients (MPCs) and MPOA servers (MPSs) dynamically learn of each other's existence. MPCs and MPSs discover each other by the exchange of messages. These messages carry the MPOA device type (MPC or MPS) and its ATM address.

3. *Target resolution.* This operation uses a modified NHRP Resolution Request message to enable MPCs to find an appropriate ATM node to reach an IP destination (target) address. This ATM node is known as a shortcut to the destination.

4. *Connection management.* This operation controls the ongoing management of ATM virtual circuits.

5. *Data transfer.* This operation is responsible for forwarding of IP traffic across a shortcut.

MPOA incorporates the use of LANE and NHRP. Figure 3–9 is an example of the contents of the NHRP Request and Reply messages. The job of NHRP in this example is to map IP addresses to ATM addresses. To keep matters simple, we use the letters DEF and KLM to represent ATM addresses, which are based on the ITU-T Network Service Access Point (NSAP) standard.

We assume the client with addresses DEF/192.168.3.3 receives a datagram destined for station 192.168.2.3 that is located in NBMA 2.[4] Client DEF sends an NHRP request message to the server. Notice that

---

[4]NBMA are the initials for a nonbroadcast multiple access network, such as X.25 or Frame Relay, that is, a switched network.

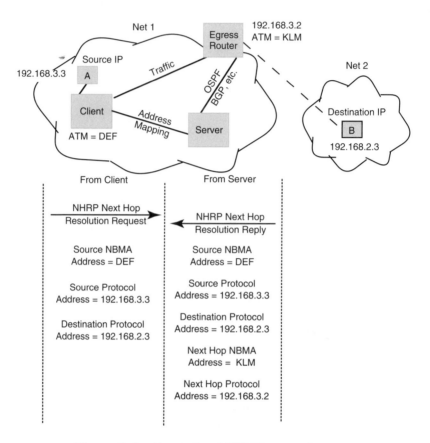

Figure 3–9   Example of NHRP operations.

the request message has the destination protocol address coded as the "target" protocol address of 192.168.2.3.

The NHS NHRP tables reveal that address 192.168.2 is reachable through KLM/192.168.3.2. Therefore, the server sends back the NHRP response message with the next hop fields coded to identify the egress router to NBMA 2. This router is identified with address KLM/192.168.3.2.

TAG SWITCHING

As mentioned earlier, tag switching is a form of label switching. The concepts are based largely on Cisco's efforts. Tag switching is based on the use of a label (a tag) in place of an address for the relaying decision; it is described in RFC 2105. The ATM Forum's Multiprotocol Over ATM is one example of a tag switching specification. The efforts by Cisco have resulted in the Multiprotocol Label Switching (MPLS) group, which is working to publish a "vendor neutral" label switching protocol. Many of the concepts explained in this part of the chapter are quite similar to those of MPLS and are implementations of many of the basic concepts introduced in Chapter 2.

Tag switching consists of two components: forwarding and control. The forwarding component uses the tag information (tags) carried by packets and the tag forwarding information maintained by a tag switch to perform packet forwarding. The control component is responsible for maintaining correct tag forwarding information among a group of interconnected tag switches.

Forwarding Component

The forwarding operation employed by tag switching is based on label swapping. When a packet with a tag is received by a tag switch, the switch uses the tag as an index in its tag information base (TIB). Each entry in the TIB consists of an incoming tag[5] and one or more subentries of the form (outgoing tag, outgoing interface, outgoing link information). If the switch finds an entry with the incoming tag equal to the tag carried in the packet, then it replaces the tag in the packet with the outgoing tag, if appropriate, and replaces the link information (e.g., MAC address) in the packet with the outgoing link-level information and forwards the packet over the outgoing interface.

The forwarding decision is based on the exact match algorithm, using a fixed-length short tag as an index. This enables a simplified forwarding procedure, in comparison to longest match forwarding traditionally used at the network layer. The forwarding procedure is simple enough to allow a hardware implementation.

The forwarding decision is independent of the tag's forwarding granularity. For example, the same forwarding algorithm applies to both unicast and multicast—a unicast entry would just have a single (outgoing

---

[5]Not stated in RFC 2105 is the possibility of associating the tag with its incoming interface, thus allowing the labels to be reused at each interface. This method allows the use of fewer bits for the label, instead of using a pool of labels for all the interfaces. Also, multicasting requires the association of the incoming datagram to the incoming interface. However, as noted later in this chapter, associating the tag with the incoming interface entails additional overhead.

tag, outgoing interface, outgoing link information) subentry, while a multicast entry may have one or more (outgoing tag, outgoing interface, outgoing link information) subentries. (For multi-access links, the outgoing link-level information in this case would include a multicast MAC address.)

The forwarding procedure is decoupled from the control component of tag switching. New routing (control) functions can be deployed without disturbing the forwarding operation.

Tag Encapsulation

Tag information can be carried in a packet in a variety of ways.

- As a small "shim" tag header inserted between the layer 2 and layer 3 headers
- As part of the layer 2 header if the layer 2 header provides adequate semantics (e.g., ATM)
- As part of the header (e.g., using the flow label field in IPv6)

The tag forwarding component is L_3 independent. Use of control component(s) specific to a particular protocol enables the use of tag switching with different L_3 protocols.

Control Component

The control component creates tag bindings and then distributes the tag binding information among tag switches. The control component is organized as a collection of modules, each supporting a particular routing function. To support new routing functions, new modules can be added. The next four sections describe some of the modules.

Destination-Based Routing.   To support destination-based routing with tag switching, a tag switch (just like a router) participates in the operations of the routing protocols (e.g., OSPF, BGP) and constructs its tag forwarding information base (TFIB) with the information it receives from these protocols.

There are three methods for tag allocation and TFIB management: (a) downstream tag allocation, (b) downstream tag allocation on demand, and (c) upstream tag allocation. In all three methods, a switch allocates tags and binds them to address prefixes in its TFIB.

Tag Allocation. In the downstream allocation, the tag that is carried in a packet is generated and bound to a prefix by the switch at the downstream end of the link (with respect to the direction of data flow). In the upstream allocation, tags are allocated and bound at the upstream end of the link. On-demand allocation means that tags will only be allocated and distributed by the downstream switch when requested by the upstream switch. The last two methods are most useful in ATM networks.

The downstream tag allocation scheme operates as follows: for each route in its TFIB, the switch allocates a tag, creates an entry in its TFIB with the incoming tag set to the allocated tag, and then advertises the binding between the (incoming) tag and the route to other adjacent tag switches. When a tag switch receives tag binding information for a route and that information was originated by the next hop for that route, the switch places the tag (carried as part of the binding information) into the outgoing tag of the TFIB entry associated with the route. This creates the binding between the outgoing tag and the route.

With the scheme for downstream tag allocation on demand, for each route in its TFIB, the switch identifies the next hop for that route. It then issues a request through a Tag Distribution Protocol (TDP) to the next hop for a tag binding for that route. When the next hop receives the request, it allocates a tag, creates an entry in its TFIB with the incoming tag set to the allocated tag, and then returns the binding between the (incoming) tag and the route to the switch that sent the original request. When the switch receives the binding information, the switch creates an entry in its TFIB and sets the outgoing tag in the entry to the value received from the next hop.

The upstream tag allocation scheme is used as follows. If a tag switch has one or more point-to-point interfaces, then for each route in its TFIB whose next hop is reachable by one of these interfaces, the switch allocates a tag, creates an entry in its TFIB with the outgoing tag set to the allocated tag, and then advertises to the next hop (through TDP) the binding between the (outgoing) tag and the route. When a tag switch that is the next hop receives the tag binding information, the switch places the tag (carried as part of the binding information) into the incoming tag of the TFIB entry associated with the route.

Once a TFIB entry is populated with both incoming and outgoing tags, the tag switch can forward packets for routes bound to the tags by using the tag switching forwarding algorithm.

When a tag switch creates a binding between an outgoing tag and a route, the switch updates its TFIB with the binding information.

A tag switch will try to populate its TFIB with incoming and outgoing tags for all routes it can reach, so that all packets can be forwarded by simple label swapping. Tag allocation is thus driven by topology (routing), not traffic—it is the existence of a TFIB entry that causes tag allocations, not the arrival of data packets.

Use of tags associated with routes, rather than flows, also means that flow classification procedures need not be performed for all the flows to determine whether to assign a tag to a flow.

Multicast and Tag Switching.   Essential to multicast routing is the notion of spanning trees. Multicast routing procedures are responsible for constructing such trees (with receivers as leaves), and multicast forwarding is responsible for forwarding multicast packets along such trees.

To support a multicast forwarding function with tag switching, each tag switch associates a tag with a multicast tree as follows. When a tag switch creates a multicast forwarding entry (either for a shared or for a source-specific tree) and the list of outgoing interfaces for the entry, the switch also creates local tags (one per outgoing interface). The switch creates an entry in its TFIB and populates (outgoing tag, outgoing interface, outgoing MAC header) it with this information for each outgoing interface, placing a locally generated tag in the outgoing tag field. This creates a binding between a multicast tree and the tags. The switch then advertises, over each outgoing interface associated with the entry, the binding between the tag (associated with this interface) and the tree.

When a tag switch receives a binding between a multicast tree and a tag from another tag switch, if the other switch is the upstream neighbor (with respect to the multicast tree), the local switch places the tag carried in the binding into the incoming tag component of the TFIB entry associated with the tree.

When a set of tag switches is interconnected through a multiple-access subnetwork, the tag allocation procedure for multicast has to be coordinated among the switches. In all other cases, the tag allocation procedure for multicast could be the same as for tags used with destination-based routing.

Flexible Routing (Explicit Routes).   One of the properties of destination-based routing is that the only information from a packet that is used to forward the packet is the destination address. While this property enables scaleable routing, it also limits the ability to influence the actual paths taken by packets. This, in turn, limits the ability to evenly distribute traffic among multiple links, taking the load off highly utilized links and shifting it towards less utilized links. For Internet service providers (ISPs) who support different classes of service, destination-based routing also limits the ability to segregate different classes with respect to the links used by these classes. Some of the ISPs today use Frame Relay or ATM to overcome the limitations imposed by destination-based routing. Tag switching, because of the flexible granularity of tags, is able to overcome these limitations without using either Frame Relay or ATM.

To provide forwarding along the paths that are different from the paths determined by the destination-based routing, the control component of tag switching allows installation of tag bindings in tag switches that do not correspond to the destination-based routing paths. Of course, this idea is also fundamental to MPLS.

Tag Switching with ATM

Since tag switching is based on label swapping and since ATM forwarding is also based on label swapping, tag switching technology can readily be applied to ATM switches by implementation of the control component of tag switching.

The tag information needed for tag switching can be carried in the VCI field. If two levels of tagging are needed, then the VPI field could be used as well, although the size of the VPI field limits the size of networks in which this would be practical.[6] However, for most applications of one level of tagging, the VCI field is adequate.

To obtain the necessary control information, the switch should be able to participate as a peer in L_3 routing protocols (e.g., OSPF, BGP). If the switch has to perform routing information aggregation, then to support destination-based unicast routing, the switch should be able to perform L_3 forwarding for some fraction of the traffic as well.

Supporting the destination-based routing function with tag switching on an ATM switch may require the switch to maintain not one, but several tags associated with a route (or a group of routes with the same next hop). This is necessary to avoid the interleaving of packets that arrive from different upstream tag switches but that are sent concurrently to the same next hop. Either the downstream tag allocation on demand

---

[6]This statement from RFC 2105 is open to question. If the VPI label is terminated at each interface, its size should not present a problem.

or the upstream tag allocation scheme could be used for the tag allocation and TFIB maintenance procedures with ATM switches.

Therefore, an ATM switch can support tag switching, but at the minimum, it needs to implement L_3 routing protocols and the tag switching control component on the switch. It may also need to support some network layer forwarding.

Implementing tag switching on an ATM switch simplifies integration of ATM switches and routers—an ATM switch capable of tag switching appears as a router to an adjacent router.

Quality of Service

Two mechanisms are needed for providing QOS to packets passing through a router or a tag switch. First, packets are identified as different classes. Second, appropriate QOS (bandwidth, loss, etc.) is provided to each class.

Tag switching provides an easy way to mark packets as belonging to a particular class after they have been classified the first time. Initial classification would be done with information carried in the L_3 layer or higher-layer headers, a concept explained in this book as an FEC. A tag corresponding to the resultant class would then be applied to the packet. Tagged packets can then be handled by the tag switching routers in their paths without needing to be reclassified.

Examples of Tag Switching Operations

This section pieces together many of the concepts just discussed. Figure 3–10 serves as an initial example for this discussion. We use generic addresses in this example for simplicity. Addresses XYZ and HIJ can be IP, IPX, ATM, AppleTalk addresses, and so on. Most likely, they are subnet or aggregated addresses with address prefixes.

The TSRs perform local binding (LB) of tags to interfaces. For example, TSR A has stored in its TFIB a binding of tag 21 to interface b. The other TSRs have stored bindings as follows:

TSR B 30.a
TSR C 21.d
TSR D 14.b

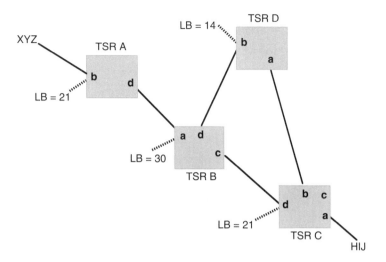

Figure 3–10   Tag switching routes (TSRs).

Through ongoing link-state route advertising (such as OSPF), routing information is flooded to the TSRs. Let us assume packets will be sent to address HIJ, so OSPF is used to build forwarding tables in each TSR to enable the TSR to forward traffic to a next hop to reach HIJ.

When TSR B finds a router (a next hop) to HIJ, it selects a tag from a pool of free tags. It uses the tag to index into its TFIB and updates this entry with (a) incoming tag of 30, (b) the associated incoming interface of a, (c) the address of the next hop, and (d) the outgoing interface to that next hop.

An important note: Our description of this operation assumes that TSR B knows the incoming interface of 30 in relation to a binding between XYZ and HIJ. Well, it may not. If the OSPF information came from (downstream) TSR C, TSR B cannot know the upstream interface until it *actually* receives a datagram destined for HIJ on interface a. This approach of associating tags with outgoing *and* incoming interfaces requires extra bookkeeping and more steps in binding the two tags/interfaces together. A viable alternative is to not associate a tag with an incoming interface, which means the tag pool is for all interfaces. This approach works if the length (in bits) of the tag is sufficient for all interfaces. Anyway, it requires some thought, and I have shown the TFIBs in Table 3–1 and Table 3–2 both ways: with and without an association to the incoming interface. Be aware that the Cisco tag switching specification does not populate the incoming interface.

The results of the TFIB updates are shown in Table 3–1. Notice that the outgoing tag in the TFIB is not yet populated. Thus far, the local binding operations have only created information on incoming tags. The outgoing tags are populated by the TSRs distributing their local binding information, which is discussed next.

In Figure 3–11, TSR B distributes its local binding information to all TSRs in the routing domain (TSRs C and D). Both TSR C and D ignore this information *because* TSR B is not the next hop to HIJ.

In Figure 3–12, TSR B receives the binding information from TSR C and TSR D. It ignores the information for TSR D because this node is not the next hop to HIJ. It accepts the information for TSR C because C is the next hop to HIJ. This remote binding information is used to populate TSR B's TFIB by means of the tag of 21 from TSR C's message.

Eventually, the TFIBs are fully populated, as shown in Table 3–2.

Border (Edge) TSRs

Tag switching is designed to scale to large networks. Its scaling capability is based on its ability to carry more than one tag in the packet.

Table 3–1(a)   TFIB: Initial Population with Correlation to Incoming Interface

TSR	Incoming Tag	Incoming Interface	Outgoing Tag	Outgoing Interface	Next Hop to HIJ
A	21	b	—	d	TSR B
B	30	a	—	c	TSR C
C	21	d	—	a	HIJ (DIR)
D	14	b	—	a	TSR C

(Note: Row entries reflect an entry at each TSR.)

Table 3–1(b)   TFIB: Initial Population with No Correlation to Incoming Interface

TSR	Incoming Tag	Outgoing Tag	Outgoing Interface	Next Hop to HIJ
A	21	—	d	TSR B
B	30	—	c	TSR C
C	21	—	a	HIJ (DIR)
D	14	—	a	TSR C

(Note: Row entries reflect an entry at each TSR.)

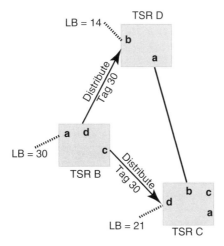

Figure 3–11   Tag information distribution.

As we will see, *tag stacking* allows designated TSRs to exchange information with each other and act as border nodes to a large domain of networks and other TSRs. These other TSRs are *interior* nodes to the domain and do not concern themselves with interdomain routes, a concept also found in MPLS.

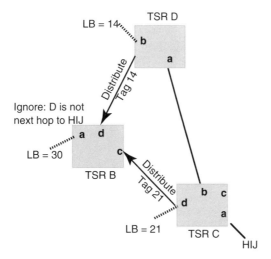

Figure 3–12   Remote building.

Table 3–2(a)   TFIB: Final Population with Correlation to Incoming Interface

TSR	Incoming Tag	Incoming Interface	Outgoing Tag	Outgoing Interface	Next Hop to HIJ
A	21	b	30	d	TSR B
B	30	a	21	c	TSR C
C	21	d	—	a	HIJ (DIR)
D	14	b	17	a	TSR C

(Note: Row entries reflect an entry at each TSR.)

Table 3–2(b)   TFIB: Final Population with No Correlation to Incoming Interface

TSR	Incoming Tag	Outgoing Tag	Outgoing Interface	Next Hop to HIJ
A	21	30	d	TSR B
B	30	21	c	TSR C
C	21	—	a	HIJ (DIR)
D	14	17	a	TSR C

(Note: Row entries reflect an entry at each TSR.)

In Figure 3–13, we modify Figure 3–10 for this discussion. Assume that three TSRs are members of the same domain (domain B) and TSR A and TSR C are border TSRs. This example also assumes that this domain is a transit domain (in which the packets traversing it neither originate nor terminate in this domain). It is certainly desirable to isolate the intradomain TSRs from these operations. In fact, we will show that the interior TSRs need to store in their TFIBs only the routing information to reach their correct border router.

Border TSR X and TSR Y are the designated border routers for domains A and C, respectively. To advertise addresses from, say, domain C, TSR Y distributes information to TSR C, which distributes it to TSR A, which then distributes it to TSR X. It is not distributed to TSR B because TSR B is an interior TSR.

We will dispense with the interface entries in Table 3–3, which shows the TFIB from domain B.

Assume the following: (a) TSR Y creates a local binding for the HIJ addresses in domain C, using 8 as the local tag; (b) TSR C creates a local binding for the same HIJ addresses, using 4 as the local tag; (c) TSR A does the same with a local tag of 7.

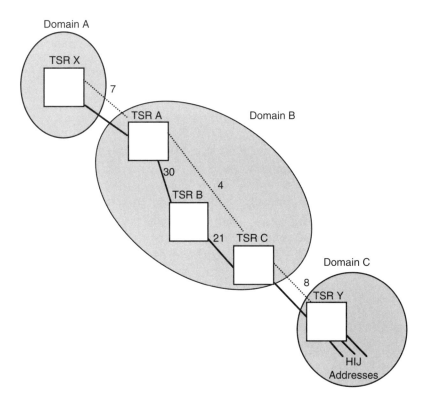

Figure 3–13   Border TSRs and domains.

Under the tag distribution procedures discussed earlier, the following events occur: (a) TSR Y distributes its tag 8 to TSR C, which uses it as an outgoing tag to the HIJ addresses; (b) TSR C distributes its tag 4 to TSR A, which uses it as an outgoing tag to the HIJ addresses; and (c) TSR A distributes its tag 7 to TSR X.

In addition, when TSR A receives the tag binding message from next hop border TSR C, it notes that TSR C is not connected directly to TSR A. It must store this information and use it later for ongoing packet transfer, discussed next.

Now, assume that a packet destined for one of the addresses in the aggregate HIJ address arrives at TSR X. TSR X sends this packet to TSR A with tag = 7. At TSR A, tag 7 is swapped for tag 4. Also, TSR A knows that this tag pertains to border TSR C, which is not connected directly to TSR A. To route the packet, TSR A pushes tag 30 onto the tag stack in the packet.

Table 3–3   TFIB for Domain B

TSR	IncomingTag	OutgoingTag	Next Hop
A	—	30	B
B	30	21	C
C	21	—	—

Therefore, the packet will contain two tags when transiting this routing domain. This approach keeps the internal TSRs isolated from interdomain routing.

To continue this example, TSR A sends the packet to TSR B, and TSR B swaps tag 30 for tag 21, then sends the packet to TSR C. TSR C's analysis of tag 21 must reveal that TSR C is to pop the tag stack in the packet, where tag 4 is found (placed there by the border TSR A). TSR C then swaps tag 4 with tag 8 and sends the packet to TSR Y and domain C, where address HIJ can be found.

Flow Classification

This book has made several comments on flow classification. It plays a role in all label switching networks. I include it in our taxonomy to make certain its importance is recognized.

Recall that a flow is a sequence of user packets from one source machine and application to one or more machines and applications. A router can cache information about the flow and circumvent the traditional IP routing mechanisms (subnet masking, search on longest subnet mask, and so on) by storing the routing information in cache, thus achieving high throughput and low delay. The flow is usually associated with an FEC, a topic explained in Chapter 2.

IPv6 Flow Operations

The IPv6 header contains a flow label field. When IPv6 is eventually implemented, this field will likely be used to support FEC and label switching operations. Figure 3–14 illustrates the format of IPv6 datagram (also called a packet, in some literature).

The *flow label* field handles different types of traffic, such as voice, video, or data. The flow label field is a special identifier that can be attached to the datagram to permit it to be given special treatment by a router. It is called the flow label field because its intent is to identify traffic in which multiple datagrams are "flowing" from a specific source

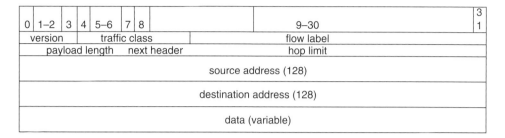

Figure 3–14   IPv6 datagram.

address to a specific destination address. The flow label field can be used in place of the IP destination address fields, but its specific use is implementation-specific. RFC 2460 provides guidance on the use of flow labels.

The *priority* field is replaced by the *traffic class* field in RFC 2460, which obsoletes the original IPv6 RFC 1883. As of the writing of the new RFC, the specific use of this field had not been defined. Notwithstanding, a number of working papers and RFCs have clarified its use with regard to other operations, such as differentiated services.

IPv4 CODEPOINT OPERATIONS

A revision to the IP TOS field has been made to support differentiated services (DS). It is called the DS codepoint (DSCP). The DS codepoint is explained in considerable detail in Chapter 12.

MPLS AND LDP

MPLS is the Internet standard for label switching. The idea of MPLS is to use an operation at the edge of the network to assign labels to each packet. These labels, in turn, are used to route the traffic from the source to the destination. Since this book concentrates on MPLS and LDP, further discussions on this part of the taxonomy are deferred to later chapters.

IPv6 Label Operations

The label-assigning operation uses the IPv6 flow label field in the header. I place this category separately from the IPv6 flow category because the label need not be associated with a flow (but in practice, it probably will be).

SUMMARY

It is obvious from this chapter that the various concepts of label switching and forwarding have many overlapping characteristics. Indeed, some of them are more similar to each other than they are different. Nonetheless, each one has its own personality and each one has different backers in the industry. How each of these technologies will fare in the market has yet to be decided. However, as of this writing, ATM-based label switching has been widely deployed, and work is complete on MPLS and nearing completion on supporting protocols, the subject of the next chapters.

4

MPLS Key Concepts

This chapter examines the Multiprotocol Label Switching (MPLS) architecture, based on RFC 3031. The major features of MPLS are explained with emphasis on the operations of LSRs and label assignments, swapping, merging, and aggregation. Label switching path (LSP) tunnels are investigated, along with label stacks and label hierarchies.

Label distribution protocols (LDPs) are responsible for the distribution of labels. Some general comments are made in this chapter about this subject, with Chapter 5 devoted to label distribution.

MAJOR ATTRIBUTES OF MPLS

Let's begin by describing some of the major attributes of MPLS, with a summary in Table 4–1. First, MPLS supports the concepts of streams (flows) and labels. The labels are not mapped end-to-end, but are managed locally between neighboring label switching routers (LSRs). Alternately, they may be passed through an LSR, unexamined, to other routers on the LSP.

MPLS may or may not use an underlying backbone technology, such as Frame Relay or ATM, and the specifications do not restrict an MPLS network in this regard. In fact, the MPLS label can reside in the DLCI or VPI/VCI fields of the Frame Relay or ATM headers, respectively.

Table 4–1   Major Attributes of MPLS

- Supports streams and labels
- Can use various L_2 networks
- Supports source (explicit) routing
- Is compatible with OSPF and BGP
- Uses labels that are local
- Uses edge device concept
- Allows QOS to be inferred from label

MPLS supports the concepts of source (or explicit) routing, wherein the originator of the traffic can dictate the route through a routing domain. However, unlike the case with conventional source routing protocols, the route need not be carried in each protocol data unit, such as the source routing option in IP. The label can be used to represent the route. MPLS does not replace, and indeed is compatible with, OSPF and BGP. Also, MPLS implements the edge device concept, wherein much of the work and processing overhead is performed before the traffic enters a core network.

MPLS eliminates the need to use the Next Hop Resolution Protocol (NHRP) and cut-through SVCs (discussed in Chapter 3), which in turn eliminates the latency associated with these operations.

The assignment of a packet to an FEC is done once, at the edge LSR, which is associated with a label. Afterward there is no further examination of the IP header. Instead, the label is used as an index into a table, which specifies the new label and the next hop for the packet.

In addition, the QOS operations performed on the packet can be (but are not required to be) inferred from the label itself. So, the label can represent the FEC and the QOS associated with the packet.

TERMINOLOGY

In addition to the terms explained in the last three chapters, these terms are used hereafter in this book.

- *Label merging.* The replacement of multiple incoming labels for a particular FEC with a single outgoing label.
- *Label switched hop.* The hop between two MPLS nodes on which forwarding is done using labels.

- *Label switched path.* The path through one or more LSRs, followed by packets in a particular FEC.
- *Label stack.* An ordered set of labels.
- *Merge point.* A node at which label merging is done.
- *Switched path.* Synonymous with label switched path.
- *VC merge.* Label merging in which the MPLS label is carried in the ATM VCI or combined VCI/VPI fields.
- *VP merge.* Label merging in which the MPLS label is carried in the ATM VPI field. The same VCI in two cells indicates that the cells originated from the same node.
- *Convergence.* The process of stabilizing IP routing tables and/or MPLS label switching tables after (a) a network startup, (b) a link or node failure, or (c) a routing change for administrative purposes.
- *Label distribution peers (peer nodes).* Two label switching routers that have set up a TCP session to exchange label/FEC binding information. These nodes are said to have a label distribution adjacency.

NETWORK MODEL FOR EXAMPLES

Figure 4–1 is used in this chapter to show examples of the MPLS operations. Each node is identified with an IP address. Six core nodes exist, and four noncore nodes are connected to the core network. The noncore nodes can be POPs, end-user routers, or MAE nodes. Some examples in this chapter show nodes A, B, G, and H as MPLS nodes, and others show them as non-MPLS nodes. The latter nodes run native-mode IP and do not have MPLS installed.

For these examples, the destination is identified with IP prefix 192.168.20.0/24 or a full address of 192.168.20.11/32.

The addresses are assigned in a simple serial, contiguous order to assist us in studying this chapter. Obviously, in the Internet, different networks and subnetworks and resulting addresses and address prefixes come into play for the end-to-end transfer of traffic.

Figure 4–2 shows another illustration that is used for the examples in this chapter. The boxes with numbers inside them represent labeled packets flowing from the upstream nodes to downstream nodes. As explained in Chapter 2, each LSR swaps the label value in the incoming packet for another value in the outgoing packet. It bears repeating that

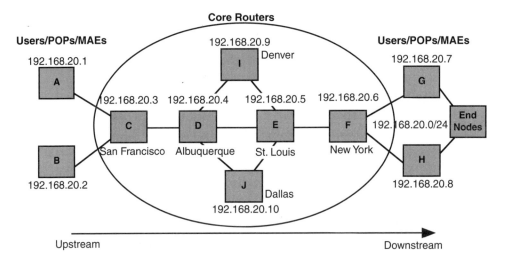

Figure 4–1   Template for examples in this chapter.

some of the examples in this chapter show nodes A, B, G, and H as non-MPLS nodes. For these examples, these nodes operate at L_3, with IP used to forward traffic.

In addition, when you read about specific instances of an actual MPLS operation, I am showing how the operations can unfold and stay in conformance with RFC 3031. Also, several examples are based on Cisco's implementation of MPLS.

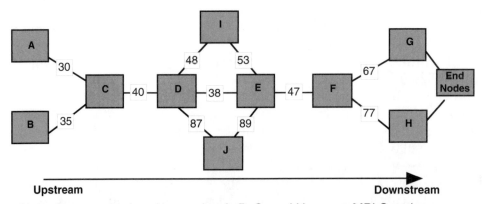

Note: Some examples show nodes A, B, G, and H as non-MPLS nodes

Figure 4–2   Labels for examples in this chapter.

TYPES OF MPLS NODES

Figure 4–3 shows the three types of MPLS nodes. They perform the following functions:

- *Ingress LSR*. Receives native-mode user traffic (for example, IP datagrams) and classifies it into an FEC. It then generates an MPLS header and assigns it a label. The IP datagram is encapsulated into the MPLS PDU, with the MPLS header attached to the datagram. If it is integrated with a QOS operation (say, DiffServ), the ingress LSR will condition the traffic in accordance with the DiffServ rules.
- *Transit LSR*. Receives the PDU and uses the MPLS header to make forwarding decisions. It will also perform label swapping. It is not concerned with processing the L_3 header, only the label header. As noted, some papers call this LSR an *interior LSR* or a *core LSR*.
- *Egress LSR*. Performs the decapsulation operations, in that it removes the MPLS header.

Figure 4–4 shows how the nodes process the labels in the packets starting at node A and going to node G. For this specific example, nodes A, B, G, and H are not configured with MPLS, so they process L_3 IP datagrams and conventional IP addresses. Node C is the ingress LSR for this MPLS routing domain, and node F is the egress LSR. Nodes D and E are transit LSRs. This example uses generic addresses to save space in the figure. For example, the mapping table at node C shows "IP=G". This

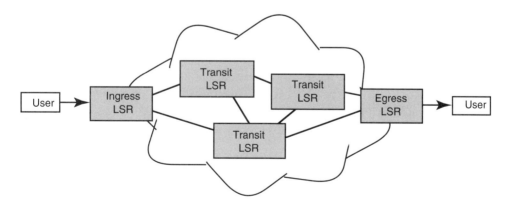

Figure 4–3   MPLS nodes.

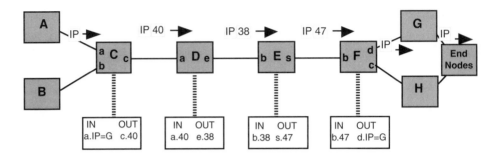

Figure 4–4   Label swapping and forwarding.

notation connotes an IP address, which as Figure 4–1 reveals, is an IP address prefix associated with an end node.

LSR C receives an IP datagram from user node A on interface a. This datagram is destined for node G. LSR C analyzes the FEC, correlates the FEC with label 40, encapsulates the datagram behind a label header, and sends the packet to output interface c. The OUT entry in LSR C's table directs it to place label 40 onto the label header in the packet. This operation at LSR C is called a label push and is explained shortly.

Hereafter, LSRs D and E process only the label header, and their swapping tables are used (at LSR D) to swap label 40 for label 38, and (at LSR E) to swap label 38 for label 47.

Notice that the swapping tables use the ingress and egress interfaces at each LSR to correlate the labels to the ingress and egress communications links. Egress LSR F is configured to recognize label 47 on interface b as the label at the end of the label switching path; that is, there are no more MPLS hops in this label switched path. The OUT entry in F's table directs LSR F to send this datagram to G on interface d; this implies removal of the label from the packet. This label removal is part of an operation that is called a label pop and is explained shortly.

UNIQUENESS OF LABELS

MPLS allows considerable leeway in how the labels are set up and used in a routing domain. The essential requirement is that a label be unambiguous with regard to identifying an FEC. This statement seems like a simple requirement, but it might not be so easy to implement. For example, assume LSR E in St. Louis receives label 6 from both LSRs D in

Albuquerque and J in Dallas. Or to take another example, a label can be received from a node that is not a direct neighbor. We explore this non-direct neighbor idea later (and it is a very powerful MPLS feature), but for now, let's assume that it can happen.

Whatever the case may be, an LSR must not bind a label to two different FECs unless it has some method to know that the packets coming in from the two different LSRs are indeed from those specific LSRs. So, while MPLS has many rules about the bindings of labels to FECs, the main idea to keep in mind as you read this and the following chapters is the following: each LSR must be able to understand and interpret its incoming labels to correct FECs. The MPLS specification will help considerably in getting this done; now let's start exploring how.

MPLS STARTUP

The first MPLS process that occurs in the router is the creation of the Label Information Base (LIB), introduced in Chapter 2. Vendors vary on how their product accomplishes this action; we use Cisco as an example in this section and also use Cisco's terminology.

At MPLS startup time the router sends out hello messages (a subject for Chapter 5) to all routers that are attached to it with communications links. For example, in Figure 4–1, the Albuquerque router sends hellos to San Francisco, Denver, St. Louis, and Dallas. This startup sets up TCP sockets between these nodes. Assuming the Label Distribution Protocol (LDP) is used, the sessions operate over TCP port 646.

ASSIGNING LABELS TO FECs

MPLS supports two methods of label assignment to an FEC. They are called independent control and ordered control.

Independent Control

With independent control, the router assigns labels to every FEC it knows. Thus, each FEC (at a minimum, each IP address prefix) has a label assigned to it. Obviously the IP routing protocols, such as OSPF, have been used previously to obtain this information, which has been placed in the IP routing table.

We could ask why a label is assigned to every IP address prefix. After all, some addresses may not be used for traffic forwarding. For example, in Figure 4–2, the IP datagrams sent from, say, node A to node G through core nodes C, D, E, and F would not need the labels assigned to nodes I and J. However, as we see later in this chapter, the independent assignment procedure leads to faster convergence time in the event a route must be changed.

Figure 4–5 shows that LSR D is informing its peer LSRs that its local label of 40 is associated with IP address prefix 192.168.20.0/24. An important idea behind this operation is that the intent of this distribution is to have node D's neighbors use label 40 when sending traffic toward node D for this prefix. Stated another way, the upstream node uses the label value assigned by the assigning downstream LSR when sending traffic to the label/prefix to that assigning LSR.

It is evident, therefore, that label 40 will be used by upstream node C for sending any IP datagrams with the destination address of 192.168.20.x to node D. However, node D will not use label 40 for the traffic to nodes I, E, and J. For example, for sending traffic to node E, node D will use the label sent to it by node E. Referring to Figure 4–2, this label value is 38.

Let me emphasize here that node D advertises label 40 for prefix 192.168.20.0/24 to all its label distribution peers. Whether these peers use this label depends on their upstream or downstream relationship with node D with regard to the destination address.

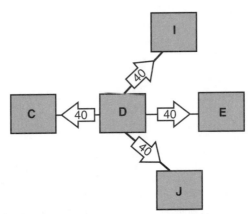

 = Binding advertisements: Label 40 with 192.168.20.0/24

Figure 4–5 Independent control.

An advantage of independent control is that the label assignment operations occur just after the advertising of the address. Assuming the address advertisement leads to rapid routing convergence (the routing tables in the routing domain are stable and are in sync with each other), the associated labels are set up quite quickly, thus allowing the network to use the more efficient labels in a timely fashion.

However, independent control should be set up so that neighbor LSRs are in agreement on the FECs (address prefixes) they will use. If the decisions are different on the FECs, some FECs may not have LSPs associated with them or they could be set up inefficiently. For example, in Figure 4–5, suppose that LSRs C and D make different choices about FECs. Both could be binding assignments as well, so the assignments might not be consistent.

Notwithstanding this potential inconsistency, in most cases it will not make any difference because of one important and simple reason: The router is interested in the labels as they pertain to the downstream flow of traffic; that is, to the next hop associated with the FEC (say, prefix 192.168.20.0/24). Thus, if node C is forwarding traffic toward node D, it will use the FEC/label advertised by D for its LFIB entry.

Ordered Control

The second method of label assignment is ordered control. It is so named because the label assignments occur in an ordered manner, either from the ingress or egress LSR of an LSP. For example, in Figure 4–1, let's assume node G discovers the nodes associated with prefix 192.168.20.0/24. Node G then initiates a label distribution that is sent to the other nodes.

Unlike independent control, ordered control ensures that all LSRs use the same FEC as the initial advertiser, LSR G in the example. This alternative also allows a network administrator some leeway in controlling how LSPs are established. For example, at the egress LSR, the administrator can configure lists that instruct the LSR as to which FECs are to be correlated with an LSP and thus are subject to label switching.

The downside to ordered control is that it takes more time than does independent control to establish the LSP. Some people consider this "latency" a small price to pay for the control it gives the network administrator. Others think ordered control is too cumbersome. For MPLS, both approaches are supported, but keep in mind that ordered control should be implemented at all LSRs if it is going to be effective. We return to the subject of ordered control in later chapters, when constrained routing is explained.

UNSOLICITED AND SOLICITED LABEL DISTRIBUTION

The independent procedure shown in Figure 4–5 is also an example of unsolicited label distribution. The LSR not only assigns but advertises the bindings to all neighbor nodes (upstream and downstream) regardless of whether these LSRs need the binding.

The other possibility is called solicited (or on-demand) label distribution. With this approach, a label binding takes place only if an LSR is requested to perform that action. Solicited label distribution is covered in later chapters.

EXAMPLE OF THE LABEL INFORMATION BASE (LIB)

Figure 4–6 shows the LIB at the Albuquerque node (node D) for the bindings it generated and for the bindings it received from its MPLS neighbors for address prefix 192.168.20.0/24. LIB entries for nonadjacent LSRs are not stored in the LIB, since they are not needed to forward packets. We also know that by the prior operations of a routing protocol, node D knows it can reach 192.168.20.0/24 through nodes I, E, and J.

Once more, the logical question is why bother having bindings with all the neighbors when some of them are not downstream with regard to the destination address and some are not even in the LSP? The answer is forthcoming shortly when in the discussion of backup routes.

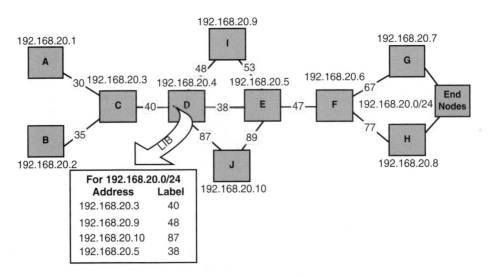

Figure 4–6   The LIB for the Albuquerque LSR (node D).

EXAMPLE OF THE LABEL FORWARDING INFORMATION BASE (LFIB)

In Figure 4–7, the LFIB at the Albuquerque LSR is shown for address prefix 192.168.20.0/24. Notice that this table contains only the information needed to forward the packet to the next hop on the LSP. Label 40 is used for this LSP between nodes C and D. Label 38 is used between nodes D and E, label 47 between nodes E and F, and so on. Label 40 is D's "local label" in that it was created by D and distributed to all of D's label distribution peers. On the other hand, label 38 was provided by node E. So, these two labels were assigned by their respective downstream nodes to their respective upstream nodes with regard to prefix 192.168.20.0/24.

The LFIB holds other information, which is not pertinent to this specific discussion. The three entries of interest in Figure 4–7 are (a) the local label for prefix 192.168.20.0/24, which is 40, (b) the outgoing label, which was assigned earlier by node E and which is 38, and (c) the physical interface for the communications link to the next hop, node E, which is n.

Two key points should be emphasized about the MPLS operations explained to this point of our analysis:

- If the LSR does not receive a label binding message from the next hop LSR, the packets will be sent unlabeled.

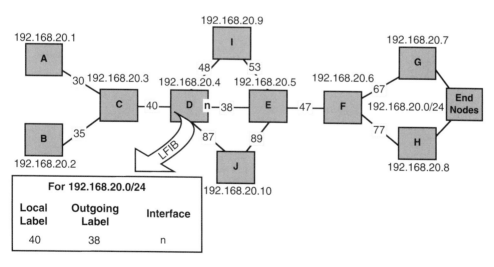

Figure 4–7   The LFIB for the Albuquerque LSR (node D).

- If the LSR does receive a label binding from the next hop LSR, this label and the local label are entered into the LFIB, as shown in Figure 4–7.

LIBERAL AND CONSERVATIVE RETENTION MODES

MPLS defines two methods for maintaining labels. The first mode is the one used thus far in this chapter. It is called liberal retention mode. As shown in Figure 4–6, label and prefix bindings are stored for upstream and downstream nodes. The other method is called conservative retention mode. With this operation, the LSR stores only the label or prefix binding assigned by the downstream LSR.

HOW MPLS CAN RECOVER FROM LINK OR NODE FAILURES

We now have enough information to examine how an MPLS network can recover from a link or node failure and reach convergence. This discussion will allow us to reexamine unsolicited distribution, independent control, and the liberal retention mode, and to discover how these MPLS features can be used to reduce the convergence delay.

Discovery of Inoperability

A link or node is discovered to be inoperable by one of two methods. With the first method, the label distribution protocol informs an LSR (say, LSR D) that a peer LSR link or node is down (say, the link between LSR D and LSR E) when LSR D receives no replies to its hello packets sent to LSR E. With the second method, the L_3 routing protocol discovers the failure with its own hello operations. The following events occur:

- Node D detects a failure on the link to node E.
- The LFIB is updated to reflect this failure. It is cleaned up so that the node D to node E link is declared down.
- The failed link interface is removed from the routing table.
- This removal activates OSPF to select an alternate link and place it in the routing table. For this example, the alternate link is the interface between Albuquerque (node D) and Dallas (node J) (and

it is possible that an alternate link has already been stored in the routing table).

- The addition of a new FEC to the L_3 routing table causes an update to node D's MPLS LFIB without having to invoke a label distribution protocol. Recall that node D had already established an MPLS peer relationship with all its neighbor nodes and had set up bindings with these nodes.

Figure 4–8 shows the reconfigured relationship between nodes D and J. The most notable aspect of this figure is that label 87 is now used by node D to forward the traffic destined for 192.168.20.0/24. Thus, the value of 87 for any traffic associated with 192.168.20.0/24 is placed in all packets that are sent to node J, which is the new member of the LSP.

Consequently, the LFIB now shows label 87 associated with interface m. This relationship was created as a result of the link failure and the invocation of OSPF to find an alternate route.

What about label 40 (see Figure 4–5)? It is the label sent to node J from node D during the startup operations discussed earlier. Since this label is upstream with regard to 192.168.20.0/24, it is ignored by node J.

The value of the liberal retention mode can be seen in these examples. Since the LIB of node D already had information on label 87/destination prefix 192.168.20.0/24 with regard to downstream node

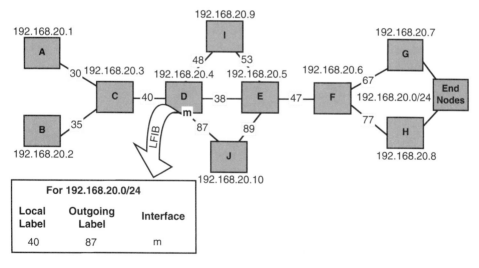

Figure 4–8   Backup and recovery.

J, it is an easy matter to update the LFIB, all without the need to invoke a label distribution protocol.

Problem with the Topology in the Network Model

The topology shown thus far in this chapter works well enough to recover from link failures, but a failure to the node itself will result in the loss of the end-to-end LSP. For example, if node E goes down, there is no way for node D to go around the failure to reach node F and the other nodes to the right side of node E. Granted, a complete failure to a node should be rare, but it does occur.

To handle this situation, in a backbone network (the core network) that supports many users, such as parts of the Internet, the prudent approach is to configure a mesh topology between the backbone nodes, as shown in Figure 4–9. With this arrangement, the core network can recover from either failed links or failed nodes. MPLS meshed networks are explained in Chapter 7. Additionally, if you would like to read in considerable detail the operations of one of the world's most famous meshed networks, take a look at Chapter 9 of [BLAC97].

THE MPLS HEADER

MPLS defines a header. It is 32 bits long and is created at the ingress LSR. It must reside behind any L_2 headers and in front of an L_3 header (IP in the example in Figure 4–10). As discussed earlier, the IP (and maybe the L_4 header) is used by the ingress LSR to ascertain an FEC,

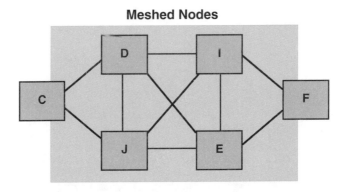

Figure 4–9   Meshing some of the LSRs.

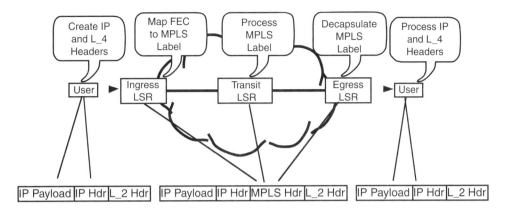

Figure 4–10   Creating and processing the MPLS header.

which in turn is used to create the label. Thereafter, and once again, only the label is processed by the transit LSRs.

The format for the MPLS header is shown in Figure 4–11. It consists of the following fields:

- *Label.* Label value, 20 bits (0–19). This value contains the MPLS label.
- *Exp.* Experimental use, 3 bits (20–22). This field is not yet fully defined. Several Internet working papers on DiffServ discuss its use with this specification.
- *S.* Stacking bit, 1 bit (23). This bit is used to indicate label stacks as discussed in the next section of this chapter.
- *TTL.* Time to live, 8 bits (24–31). Places a limit on how many hops the MPLS packet can traverse. This limit is needed because the IP TTL field is not examined by the transit LSRs.

THE LABEL STACK

Label switching is designed to scale to large networks, and MPLS supports label switching with hierarchical operations; this support is based on the ability of MPLS to carry more than one label in the packet. Label stacking allows designated LSRs to exchange information with one another and act as border nodes to a large domain of networks and other LSRs. Recall that these other LSRs are interior nodes (transit nodes) to

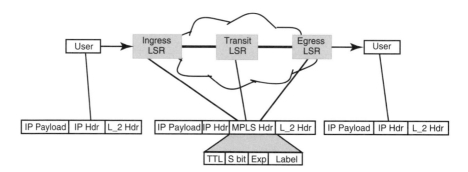

Figure 4–11   Format of the header.

the domain and do not concern themselves with interdomain routes or with the labels associated with these routes.

The processing of a labeled packet is completely independent of the level of hierarchy; that is, the level of the label is not relevant to the LSR. To keep the operations simple, the processing is always based on the top label, without regard to the possibility that some number of other labels may have been above it in the past or that some number of other labels may be below it at present. In Figure 4–4, the label stack is 1 because the packet was sent from C to D to E to F. An unlabeled packet can be thought of as a packet whose label stack is empty (i.e., whose label stack has depth 0). In Figure 4–4, the path from A to C and the path from F to G have an empty label stack, of course.

If a packet's label stack is of depth m, the label at the bottom of the stack is considered as the level 1 label, to the label above it (if such exists) as the level 2 label, and to the label at the top of the stack as the level m label. In Figure 4–12, assume that three LSRs are members of the same domain (domain B) and LSR A and LSR C are border LSRs. This example also assumes that this domain is a transit domain (in which the packets traversing it neither originate nor terminate in this domain). It is certainly desirable to isolate the intradomain LSRs from these operations.

LSR X and LSR Y are the designated border routers for domains A and C, respectively. To advertise addresses from, say, domain C, LSR Y distributes information to LSR C, which distributes it to LSR A, which then distributes it to LSR X. It is not distributed to LSR B because LSR B is an interior LSR.

Two levels of labels are used. When traffic traverses through domain B, one level of labels is used and the labels pertaining to the interdomain operations are pushed down in a label stack in the packet.

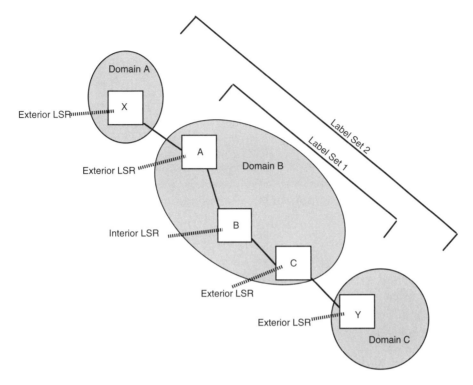

Figure 4–12   Label stacks and hierarchies.

Figure 4–13 shows examples of label stacks. Nodes A, B, G, and H are exterior nodes (ingress and egress LSRs) to the internal domain where nodes C, D, E, and F reside. The LSR tables at nodes C and F have label stacks to a depth of 2. LSR D and LSR E tables have label stacks to a depth of 1. Notice that this example is slightly different from Figure 4–4. In this example, the MPLS capabilities are extended out to nodes A, B, G, and H. Therefore, sitting behind these nodes are most likely some non-MPLS nodes, such as workstations and servers.

Node A sends a packet to node C with a label of 21. Node C consults its label table and determines that the label is to be pushed down and that label 33 is to be used between node C and node D. The packet sent to node D has two labels, but label 21 is not examined by node D. Its label table directs it to swap label 33 for 14 and relay the packet out of interface e, a link to node E.

Upon node E receiving this packet, its label table instructs it to pop up the next label and then send the packet to interface S. There is now

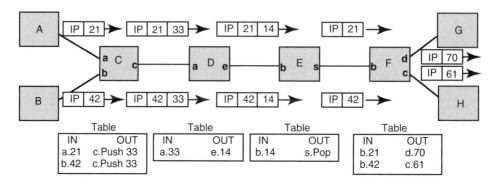

Figure 4–13   Example of label stacking: LSR pops the stack.

only one label in the header. At node F, the label value of 21 on interface b is correlated to label 70 on interface d, the link to node G.

The second example in Figure 4–13 is a packet emanating from node B, with a label value of 42. The label table at node C indicates that this label is to be pushed, and label 33 is used as the outer label. The process then proceeds in the same manner as the first example until the traffic reaches node F. Here, the pop-up operation reveals label 42, which is correlated with label 61 on interface c, the link to node H.

In this example, only one label binding was needed at the interior LSRs to handle two external labels. Of course, it is possible to map thousands of labels from exterior nodes to one label binding in an interior domain.

Figure 4–14 shows another example. In this operation, LSR F pops the stack and LSR E does not. LSR E processes the outer label just as LSR D has done.

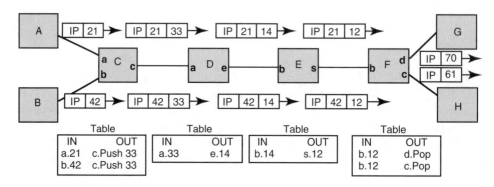

Figure 4–14   Example of label stacking: LSR F pops the stack.

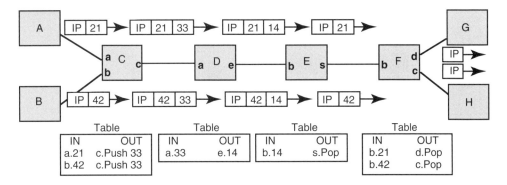

Figure 4–15   Two label pops.

Figure 4–15 shows one more example of label stacking. In this situation, nodes G and H are not LSRs. They are end stations, such as servers or routers, that are not configured for MPLS operations. Two label pops occur, the first at LSR E and the second at LSR F. These three scenarios in Figures 4–13, 4–14, and 4–15 are all permitted; the reasons for these choices are explained shortly.

STACKING RULES FOR THE LABEL SWITCHED PATH

The explanation in the sidebar explains the rules for the label stacks, as provided in Section 3.15 of the MPLS RFC 3031 [ROSE01c]. In summary, these rules ensure that when an LSR pushes a label onto a packet that is already labeled, the new label corresponds to an FEC whose LSP egress is the LSR that assigned the label that is now second in the stack.

Penultimate Hop Popping

In Figure 4–15, one label pop is performed by LSR E and the other by LSR F. One might wonder why the stacking did not continue to LSR F. The answer is that stacking can indeed be extended to LSR F. The decision as to where the final pop occurs is based on (a) the capability of the LSR and (b) the possible desire to execute no more than one label pop at each LSR.

Penultimate hop popping might be used to permit a node to perform only one label pop. The possible requirement of a node to perform more than one pop may be beyond the capability of the node, it may complicate the node's operations, or it may lead to unacceptable delay at the node.

RFC 3031 Rules for Label Stacks

A "Label Switched Path (LSP) of level m" for a particular packet P is a sequence of routers with the following properties:

1. R1, the "LSP Ingress", is an LSR which pushes a label onto P's label stack, resulting in a label stack of depth m;
2. For all i, 1<i<n, P has a label stack of depth m when received by LSR Ri;
3. At no time during P's transit from R1 to R[n-1] does its label stack ever have a depth of less than m;
4. For all i, 1<i<n: Ri transmits P to R[i+1] by means of MPLS, i.e., by using the label at the top of the label stack (the level m label) as an index into an incoming label map (ILM);
5. For all i, 1<i<n: if a system S receives and forwards P after P is transmitted by Ri but before P is received by R[i+1] (e.g., Ri and R[i+1] might be connected via a switched data link subnetwork, and S might be one of the data link switches), then S's forwarding decision is not based on the level m label, or on the network layer header. This may be because:

a) the decision is not based on the label stack or the network layer;

b) the decision is based on a label stack on which additional labels have been pushed (i.e., on a level m+k label, where k>0).

The idea is that by using penultimate label popping, the performance at each node and the end-to-end LSP is easier to calculate.

One more point about this operation: The penultimate node pops the label stack only if this process is requested by the egress node or if the next node in the LSP does not support MPLS. In the Cisco router, the egress LSR makes this request to its neighbor with label value 3, the Implicit Null Label.

Stacks and Encapsulations

The procedures for identifying the label stack and encapsulating it in lower-layer protocols, such as PPP or ATM, are specified in [ROSE01a]. The format for the encapsulated label stack was introduced in Figure 4–11.

The label stack entries appear after the data link layer headers but before any network layer headers. The top of the label stack appears earliest in the packet, and the bottom appears latest. The network layer packet immediately follows the label stack entry that has the S bit set.

When a labeled packet is received, the label value at the top of the stack is looked up. As a result of a successful lookup, the LSR knows the next hop for the packet and the operation that is to be performed on the packet, such as pushing or popping operations.

Reserved Label Values

Several label values are reserved. They are as follows:

- A value of 0 represents the "IPv4 Explicit NULL Label." This label value is only legal when it is the sole label stack entry. It indicates that the label stack must be popped, and the forwarding of the packet must then be based on the IPv4 header.
- A value of 1 represents the "Router Alert Label." This label value is legal anywhere in the label stack except at the bottom. When a received packet contains this label value at the top of the label stack, it is delivered to a local software module for processing.
- A value of 2 represents the "IPv6 Explicit NULL Label." This label value is only legal when it is the sole label stack entry. It indicates that the label stack must be popped, and the forwarding of the packet must then be based on the IPv6 header.
- A value of 3 represents the "Implicit NULL Label." This is a label that an LSR may assign and distribute but which never actually appears in the encapsulation. When an LSR would otherwise replace the label at the top of the stack with a new label but the new label is "Implicit NULL," the LSR will pop the stack instead of doing the replacement. Although this value may never appear in the encapsulation, it needs to be specified in the Label Distribution Protocol, so a value is reserved.
- Values 4 through 15 are reserved.

MPLS TERMS FOR LIB AND LFIB

The terms LIB and LFIB are used by Cisco, and they have been used thus far in this chapter because they are examples of actual MPLS label and cross-connect fabrics. The MPLS specification in RFC 3031 uses three other terms for the LIB and LFIB. They are illustrated in Figure 4–16 and described in this section.

NHLFEs

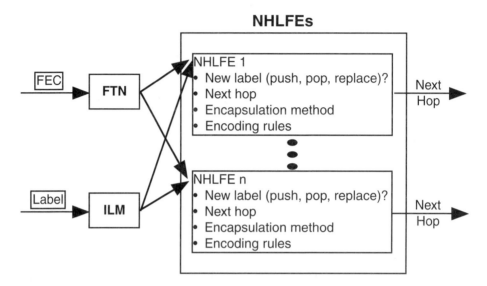

Figure 4–16   NHLFE, FTN, and ILM.

The Next Hop Label Forwarding Entry (NHLFE)

The Next Hop Label Forwarding Entry (NHLFE) is used when a labeled packet is forwarded. It contains the following information:

- The packet's next hop.
- The operation to perform on the packet's label stack; this operation is one of the following operations:

  a) Replace the label at the top of the label stack with a specified new label.

  b) Pop the label stack.

  c) Replace the label at the top of the label stack with a specified new label and then push one or more specified new labels onto the label stack.

The NHLFE can also contain information about the data link encapsulation method and about the way to encode the label stack. Both of these topics are explained in Chapter 6.

FEC-to-NHLFE Map (FTN)

The FEC-to-NHLFE (FTN) Map correlates each FEC to a set of NHLFEs. It is used when forwarding packets arrive unlabeled but are to be labeled before being forwarded. If the FTN maps a particular label to a set of NHLFEs that contains more than one element, exactly one element of the set must be chosen before the packet is forwarded.

Incoming Label Map (ILM)

The Incoming Label Map (ILM) maps each incoming label to a set of NHLFEs. It is used when forwarding packets arrive as labeled packets. If the ILM maps a particular label to a set of NHLFEs that contains more than one element, exactly one element of the set must be chosen before the packet is forwarded. The label at the top of the stack is used as an index into the ILM. Having the ILM map a label to a set containing more than one NHLFE may be useful, for example, to load-balance the traffic across multiple links.

AGGREGATION

One way of dividing traffic into FECs is to create a separate FEC for each address prefix that appears in the routing table, as shown in Figure 4–17(a). This approach may result in a set of FECs that follow the same route to the egress node, and label swapping might be used only to get the traffic to this node. In this situation, within the MPLS domain, these separate FECs do no good. In the MPLS view, the union of those FECs is itself an FEC. This situation creates a choice: Bind a distinct label to an FEC, or bind a label to the union and apply the associated label to all traffic in the union, shown in Figure 4–17(b).

The procedure of binding a single label to a union of FECs, which is itself an FEC (within the same MPLS domain), and of applying that label to all traffic in the union is known as *aggregation*. Aggregation can reduce the number of labels needed to handle a particular set of packets and can also reduce the amount of label distribution control traffic needed.

A set of FECs that can be aggregated into a single FEC can be (a) aggregated into a single FEC, (b) aggregated into a set of FECs, or (c) not aggregated at all. The MPLS specification speaks of the "granularity" of aggregation, with (a) being the coarsest granularity, and (b) being the finest granularity. The MPLS specification provides a number of rules on aggregation; see Section 3.20 of [ROSE01c].

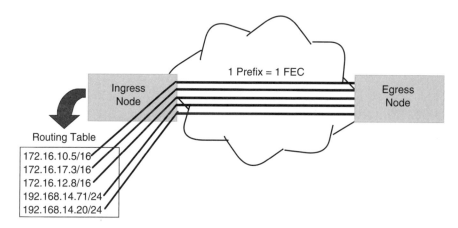

(a) Separate FEC for each address prefix

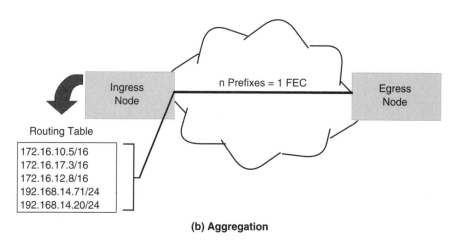

(b) Aggregation

Figure 4–17   Nonaggregation and aggregation.

LABEL MERGING

With label merging, multiple packets arriving with different labels have a single label applied to them on their outgoing interface (the same interface). The idea is illustrated in Figure 4–18. LSR C sends three packets to LSR D, with labels 21, 24, and 44 in the label header. LSR D merges these labels into label 14 and sends the three packets to LSR E.

MPLS supports LSRs that have either nonmerging or merging operations. The basic rules for these two types of LSRs is quite simple: (a) an upstream LSR that supports label merging need only be sent one label per FEC; (b) an upstream LSR that does not support label merging must

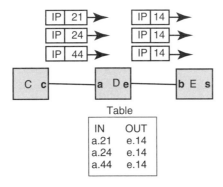

Figure 4–18   Label merging.

be sent a label for each FEC; (c) if an upstream LSR does not support label merging, it must ask for a label for an FEC.

Many of the issues surrounding label merging deal with running MPLS on ATM networks. Therefore, we defer the discussion of this aspect of label merging to Chapter 6. In addition, there are many detailed rules for how label distribution is performed when LSRs are using label merging operations. If you need each rule of these operations, refer to Sections 5.2.1 and 5.2.2 of RFC 3031.

SCOPE AND UNIQUENESS OF LABELS IN A LABEL SPACE

In Chapter 2, the subject of label space was introduced (see Figure 2–7); Figure 4–19 is another rendition of the Chapter 2 illustration. It shows that labels can be assigned across all interfaces or to each interface at an LSR.

Given the concepts, Figure 4–20 provides four scenarios of how MPLS sets the rules for the scope and uniqueness of labels. For these

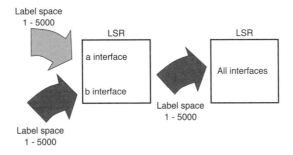

Figure 4–19   Label space reviewed.

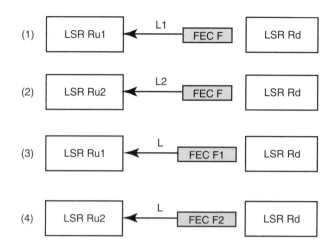

Figure 4–20   Label space and label scope and uniqueness.

examples, MPLS uses the shorthand notation Ru and Rd for LSR upstream and LSR downstream, respectively.

- Scenario 1: LSR Rd binds label L_1 to FEC F and distributes the binding to peer LSR Ru1.
- Scenario 2: LSR Rd binds label L_2 to FEC F and distributes the binding to peer LSR Ru2.
- Scenario 3: LSR Rd binds label L to FEC F1 and distributes the binding to peer LSR Ru1.
- Scenario 4: LSR Rd binds label L to FEC F2 and distributes the binding to peer LSR Ru2.

For scenarios 1 and 2, it is a local matter whether L_1 equals L_2. For scenarios 3 and 4, the following rule applies: If when Rd receives a packet whose top label is L, Rd can determine whether the label was put there by Ru1 or Ru2, then MPLS does not require that F1 equal F2. Therefore, for scenarios 3 and 4, Rd is using different label spaces for its distributions to Ru1 and Ru2, an example of the per-interface label space.

HOP-BY-HOP AND EXPLICIT ROUTING

MPLS uses two methods for choosing the LSP for an FEC; that is, for the route selection. One method is hop-by-hop routing, wherein each node independently chooses the next hop for an FEC. This approach is the practice used today in most internets, with a routing protocol such as OSPF. The other method is explicit routing. No node is allowed to choose the next hop. Instead, a designated LSR, usually the LSP ingress or egress node, specifies the LSRs that are to be in the LSP.

Two modes are permitted with explicit routing. If the entire LSP is specified, the LSP is "strictly explicitly routed." If part of the LSP is specified, the LSP is said to be "loosely explicitly routed." These modes are similar to IP's strict and loose routing options.

Explicit routing can be a valuable tool to support traffic, such as voice and video packets, that has QOS requirements. It also can play a useful role in traffic engineering operations, a subject covered in Chapter 7.

REVIEW OF THE LABEL RETENTION MODE

To summarize the label retention modes, the MPLS specifications set forth these methods for retaining or discarding a label.

- An LSR Ru may receive a label binding for a particular FEC from an LSR Rd, even though this Rd is not Ru's next hop (or is no longer Ru's next hop) for that FEC.
- Ru then has the choice of whether to keep track of such bindings, or whether to discard such bindings.
- If Ru keeps track of these bindings, it may begin using the binding again if Rd eventually becomes its next hop for the FEC in question. If Ru discards such bindings, then if Rd later becomes the next hop, the binding will have to be reacquired.
- If an LSR supports "liberal label retention mode," it maintains the bindings between a label and an FEC that are received from LSRs that are not its next hop for that FEC. If an LSR supports "conservative label retention mode," it discards such bindings.

ADVERTISING AND USING LABELS

Several label distribution mechanisms are used to advertise and distribute labels; this subject is covered in more detail in Chapter 5. The MPLS architecture specification establishes the overall procedures for these operations; this part of the chapter provides a general description of the procedures along with several illustrations to help your analysis. Also, references are made to some of Cisco's MPLS implementations described earlier in this chapter and in Chapter 6. The procedures are organized as follows.

The downstream LSR has five procedures defined, classified as four distribution procedures and one withdrawal procedure:

- Distribution procedures: (a) PushUnconditional, (b) PushConditional, (c) PulledUnconditional, (d) PulledConditional
- Withdrawal procedure

The upstream LSR has nine procedures defined, classifed by one of four names (a) request, (b) not available, (c) release, and (d) label use:

- Request procedures: (a) RequestNever, (b) RequestWhenNeeded, (c) RequestOnRequest
- NotAvailable procedures: (a) RequestRetry, (b) RequestNoRetry
- Release procedures: (a) ReleaseOnChange, (b) NoRelease-OnChange
- LabelUse procedures: (a) UseImmediate, (b) UseIfLoopNot-Detected

These operations take place though combinations of the control modes and downstream/upstream concepts explained in the first part of the chapter. Recall that they are:

- DOD   Downstream on demand
- ICM   Independent control mode
- OCM   Ordered control mode
- UDS   Unsolicited downstream

The Downstream LSR (Rd) Procedures

Figure 4–21 illustrates the procedures described in this subsection.

PushUnconditional

The PushUnconditional operation is an example of unsolicited downstream label assignment in the independent LSP control mode. It is the procedure used in Cisco routers.

In Figure 4–21(a), X is an address prefix in Rd's routing table, and Ru is a distribution peer of Rd with respect to X. When these situations hold, Rd must bind a label to X and send the binding to Ru. It then

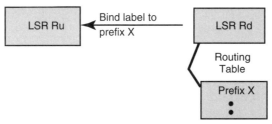

(a) PushUnconditional

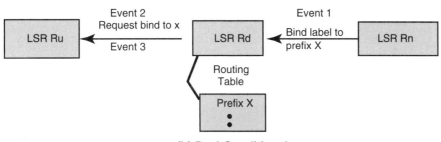

(b) PushConditional

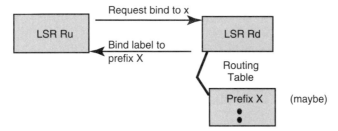

(c) PulledUnconditional

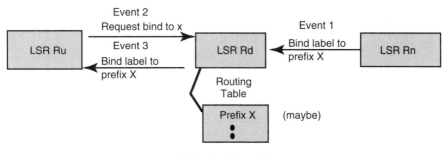

(d) PulledConditional

Figure 4–21   Downstream LSR procedures.

becomes the responsibility of Ru to keep this binding up-to-date and to keep Rd informed of any changes.

PushConditional

The PushConditional operation is shown in Figure 4–21(b). It is an example of unsolicited downstream label assignment in the ordered LSP mode. X is an address prefix in Rd's routing table. Rd is an LSP egress for X; Rd's next hop for X is Rn. Furthermore, Rn has bound a label to X and distributed that binding to Rd. In this situation, Rd should bind a label to X and send it to Ru.

The difference between PushUnconditional and PushConditional is this: PushUnconditional causes the distribution of bindings for all address prefixes in the routing table; PushConditional causes the distribution only for those address prefixes for which the Rd has received bindings from the LSP next hop.

PulledUnconditional

The PulledUnconditional operation, shown in Figure 4–21(c) is an example of downstream-on-demand label distribution, using independent LSP control mode. X is an address prefix in Rd's routing table, and Ru has explicitly requested that Rd bind a label to X and distribute the binding to Ru. Rd must honor this bind request, and if it cannot (if it is not a label distribution peer to Ru, for example), it must so inform Ru. If Rd has already sent a binding for this prefix, it must still honor this request and send a new binding. The former binding remains in effect. The end result of this operation is two labels associated with the same prefix. Why would the network operator want to do this? MPLS is terse about this rule. One thought comes to mind: If other aspects of an FEC besides the prefix are considered, it would allow the binding of different labels to different FECs with the same prefix.

PulledConditional

The PulledConditional operation, shown in Figure 4–21(d), is an example of downstream-on-demand label distribution, using ordered LSP control mode. Again, X is an address prefix in Rd's routing table, and Ru has requested that Rd bind a label to X and distribute the binding to Ru. Rd is an LSP egress for X, or Rd's L3 next hop for X is Rn and Rn has bound a label to X and distributed that binding to Rd. If these conditions hold, Rd must bind a label to X and distribute that binding to Ru.

Withdrawal Procedure

This procedure is straightforward. If an LSR decides to break a binding between a label and a prefix, the LDP unbinding message must be distributed to all LSRs to which the initial binding was distributed.

The Upstream LSR (Ru) Procedure

The upstream LSR operations are simpler than those found at the downstream LSR. They are summarized here, based on Section 5.1.2 of RFC 3031.

RequestNever

An LSR never requests a label binding. For example, in Figures 4–21(a) and (b), the downstream LSR takes the necessary actions to bind the labels to the prefixes. It is not necessary to burden the upstream LSR with these tasks. This procedure is applicable when an LSR uses unsolicited downstream label distribution and liberal label retention modes, but should not be used if the Rd uses PulledUnconditional or PulledConditional procedures.

RequestWhenNeeded

When a router finds a new prefix or when one is updated, the procedure is executed—if a label binding does not already exist. This procedure is executed by an LSR if conservative label retention mode is being used.

RequestOnRequest

This operation issues a request whenever a request is received, in addition to issuing a request when needed. If the Ru is not capable of being an LSP ingress, it may issue a request only when it receives a request from upstream. If the Rd receives such a request from Ru for an address prefix for which the Rd has already distributed a label to Ru, then Rd assigns a new label, binds it to X, and distributes that binding.

NotAvailable Procedure

The NotAvailable procedure determines how Ru responds to the following situation:

1. The Ru and Rd are, respectively, upstream and downstream label distribution peers for address prefix X.
2. Rd is Ru's L_3 next hop for X.
3. Ru requests a binding for X from Rd.
4. Rd replies that it cannot provide a binding at this time because it has no next hop for X.

Two possible procedures govern Ru's behavior: RequestRetry and RequestNoTry.

RequestRetry

The Ru should issue the request again at a later time. This procedure would be used when downstream-on-demand label distribution is used.

RequestNoRetry

The Ru should never reissue the request, instead assuming that the Rd will provide the binding automatically when it is available. This is useful if the Rd uses the PushUnconditional procedure; that is, if unsolicited downstream label distribution is used.

Release Procedure

Label release procedures simply mean that the binding of a label to an FEC is deleted at an LSR. The scenario for the release is as follows. Rd is an LSR that has bound a label to address prefix X; it has already distributed that binding to LSR Ru. If Rd is not Ru's L_3 next hop for address prefix X or has ceased to be Ru's L_3 next hop for address prefix X, then Ru will not be using the label, and it makes no sense to retain it unless there is a likelihood of this association recurring. Two possible procedures govern Ru's behavior: ReleaseOnChange and NoReleaseOnChange.

ReleaseOnChange

Ru releases the binding and informs Rd that it has done so. This procedure is used to implement conservative label retention mode.

NoReleaseOnChange

Ru maintains the binding so that it can use it again immediately if Rd later becomes Ru's L_3 next hop for X. This procedure is used to implement liberal label retention mode.

Label Use Procedure

Let us assume that Ru has received a label binding L for address prefix X from LSR Rd, and Ru is upstream of Rd with respect to X, and Rd is Ru's L_3 next hop for X. Ru will make use of the binding if Rd is Ru's L_3 next hop for X. If, at the time the binding is received by Ru, Rd is not Ru's L_3 next hop for X, Ru does not make any use of the binding at that time. Ru may, however, start using the binding at some later time if Rd becomes Ru's L_3 next hop for X. Ru can use two procedures: UseImmediate and UseIfLoopNotDetected.

UseImmediate

Ru may put the binding into use immediately. At any time when Ru has a binding for X from Rd and Rd is Ru's L_3 next hop for X, Rd will also be Ru's LSP next hop for X. This procedure is used when loop detection is not in use.

UseIfLoopNotDetected

This procedure is the same as UseImmediate, unless Ru has detected a loop in the LSP. If a loop has been detected, Ru will discontinue the use of label L for forwarding packets to Rd. This procedure is used when loop detection is in use and will continue until the next hop for X changes or until the loop is no longer detected.

SUMMARY

We have examined the major features of the MPLS specification as defined in RFC 3031. The major features of MPLS have been explained with emphasis on the operations of LSRs and label assignments, swapping, merging, and aggregation. LSP tunnels were investigated, along with label stacks and label hierarchies. We now turn our attention to methods for distributing the labels to the LSRs.

5

Label Distribution Operations

Several methods can be employed to advertise and distribute labels. This chapter examines three methods: the Label Distribution Protocol (LDP), the Resource Reservation Protocol (RSVP), and the Border Gateway Protocol (BGP). The bulk of the material in this chapter is devoted to LDP. This slant does not mean that the other two methods are not viable options. Indeed, RSVP and BGP (with revisions) are effective, but LDP is more complex and has more procedures, features, and messages than the other two. This fact is reflected in their descriptions.

THE ISSUE OF LABEL GRANULARITY

Previous parts of this book have explained the idea of label aggregation, where multiple (perhaps many) FECs are aggregated into one or a few labels. One of the central ideas of MPLS is to be able to aggregate FECs (and specifically IP destination prefixes) into one label. In this manner, the MPLS network has fewer entries in its label cross-connect table (the LFIB) and can operate more efficiently.

The potential disadvantage of the aggregation approach is that it will work against the important issue of supporting (and charging for) tailored QOS for each user (or a subset of users), who is/are identified with a unique label. After all, aggregating multiple users with different QOS into one label defeats one of the goals of the emerging new Internet: providing tailored QOS for individual users.

For the long run, I do not hold this view. When the Internet has evolved to a multiterabit transport system (which is underway now), the backbone network's performance will be of such high capacity that a backbone node will not have the time to tailor its behavior to meet each user's performance requirements. Nor should it have to do so; the network will be operating at such a high capacity and providing such low delay and high throughput that the issue of QOS support will only be addressed at the edge nodes of the network—those that provide ingress and egress for the customers' traffic.

Then why even bother with QOS provisioning and granular MPLS label management at all? If the network is so all-powerful, why not just eliminate the time-consuming QOS and traffic policing functions associated with MPLS labels?

The answer is twofold. First, QOS and traffic policing will remain essential for the network provider to charge for its services and for traffic to be regulated at the (probable) bandwidth-constrained link from the user to the network, the so-called user-network interface (UNI). Second, the idea of a backbone Internet that has all the capacity it needs is still just that: an idea.

Nonetheless, the idea of granularity in an MPLS-based backbone network will evolve to one of using just the destination IP prefix as the FEC through the backbone but resorting to a finer granularity at the UNI in order to price the QOS services furnished (or not furnished) to the user. In my view, the emerging all-optical network will obviate MPLS fine granularity operations within the core of the network. Therefore, MPLS label aggregation will continue to be an important idea, but it will be pushed out to the network edge (actually, in many networks, if it even exists, that is where it is now).

With these general thoughts in mind, let's now take a look at several label distribution protocols.

METHODS FOR LABEL DISTRIBUTION

MPLS does not stipulate a specific label distribution protocol.[1] Since several protocols that can support label distribution are currently in operation, it makes sense to use what is available. Notwithstanding, the IETF

---

[1]Some papers call a label distribution protocol a signaling protocol. If this term is used, be aware that it does not refer to conventional signaling protocols, such as ISDN's Q.931 and SS7's ISUP.

has developed a specific protocol to complement MPLS. It is called the Label Distribution Protocol.

Another protocol, the constraint-based LDP (CR-LDP), allows the network manager to set up explicitly routed label switched paths (LSPs). CR-LDP is an extension to LDP. It operates independently of any Internal Gateway Protocol (IGP). It is used for delay-sensitive traffic and emulates a circuit-switched network.

RSVP can also be used for label distribution. By using the RSVP PATH and Reservation messages (with extensions), it supports label binding and distribution operations.

BGP is a good candidate for the label distribution protocol. If it is desirable to bind labels to address prefixes, then BGP might be used. A BGP reflector can be used to distribute the labels.

ADVERTISING AND DISTRIBUTING AGGREGATED FECs

The subject of aggregation was introduced in Chapter 4 (See "Aggregation"). We pick up this subject here and discuss in more detail some of the features requirements of aggregation.

One of the attractive aspects of MPLS is that it allows considerable flexibility and independence of action on the part of an LSR as to how certain label relationships are distributed and managed. The reason for building this capability into MPLS is the recognition that traffic traversing different networks (different ISPs, different telco networks, multiple campus networks, and so on) will encounter different implementations of label management and the association of labels with different kinds of traffic.

Ru Has Finer Granularity Than That of Rd

One aspect of this flexibility is shown in Figure 5–1(a), where Ru has finer granularity than that of Rd. Rd has delivered to Ru and Rm a binding of prefixes in a label distribution protocol message associating 10.1.0.0/12 and 10.2.0.0/12 to label L1. Ru decides to disaggregate this binding and issues two binds to Rn: 10.1.0.0/12 to L1 and 10.2.0.0/12 to L2.

One reason that Ru might take this action is to provide a finer level of granularity than is available with just the IP address prefixes, perhaps using other parts of an FEC, such as a port number, to make the binding associations.

MPLS allows this kind of independence. However, MPLS requires that the Ru keep this association transparent to the Rd when the Ru

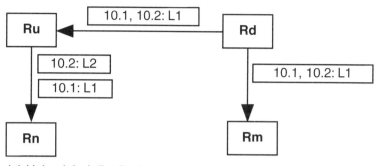

(a) Using label distribution to achieve finer granularity

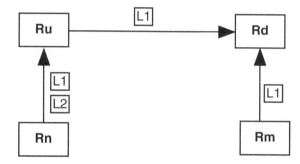

(b) Resulting aggregation operations

Figure 5–1   Aggregation and disaggregation scenarios.

sends traffic to the Rd. As shown in Figure 5–1(b), if packets containing either L1 or L2 are sent to Ru and then to Rd, the labels can be merged (aggregated) by Ru back to L1.

In essence, when the Ru needs to forward labeled packets to the Rd, it may need to map n labels into m labels, where n > m, as shown in Figure 5–1(b). As an option, the Ru may withdraw the set of n labels that it has distributed and then distribute a set of m labels, corresponding to Rd's level of granularity. In this manner, Rd is not aware of the disaggregation operations at Ru.

Ru Has Coarser Granularity Than That of Rd

It is also possible that the Ru has coarser granularity than does the Rd (Rd has distributed n labels for the set of FECs, while Ru has distributed m labels, where n > m). In this situation, RFC 3031 recommends that the Ru adopt Rd's finer level of granularity. This approach would require the Ru to withdraw m labels it has distributed and adhere to Ru's n label approach.

If this approach is not used, the Ru can map its m labels into a subset of Rd's n labels if (and a very important if) the Ru can keep the routing the same for the FECs. I don't find much attraction to this second approach; it complicates matters considerably.

INTRODUCTION TO LDP

MPLS must provide a standard method for the distribution of routing labels between neighbor LSRs. As of this writing, this standard is being defined in RFC 3036. Here we highlight the major aspects of this RFC.

We know that MPLS does not make a forwarding decision with each L_3 datagram (based on the addressing and TOS contents of the L_3 header). Instead, a forwarding equivalency is determined for classes of L_3 datagrams, and a fixed-length label is negotiated between neighboring LSRs along LSPs from ingress to egress. Routers with label switching capabilities must be able to determine which of their neighbors are capable of MPLS operations. They must then agree upon the label values to be used for the transport of user traffic. The LDP is used to support this requirement.

Figure 5–2 shows the general concepts of LDP. It operates between LSRs that are directly connected by a link (LSR A and LSR B, as well as LSR B and LSR C). It can also operate between nonadjacent LSRs: LSR A and LSR C, shown in the figure with dashed lines. Obviously, the LDP messages for the label bindings for LSRs A and C flow through LSR B, but B does not take action on them.

LSRs that use LDP to exchange label and FEC mapping information are called *LDP peers;* they exchange this information by forming an LDP session.

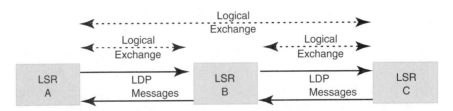

Figure 5–2   LDP message exchanges.

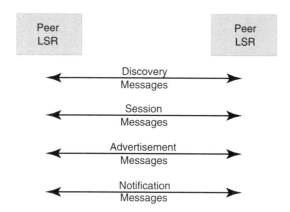

Figure 5–3   Categories of LDP messages.

LDP MESSAGES

The four categories of LDP messages are shown in Figure 5–3. *Discovery* messages announce and maintain the presence of an LSR in a network. Periodically, an LSR sends a Hello message through a UDP port with the multicast address of "all routers on this subnet."

The session messages establish, maintain, and delete sessions between LDP peers (the LSRs). These operations entail the sending of Initialization messages over TCP. After this operation is complete, the two LSRs are LDP peers.

Advertisement messages create, change, and delete label mappings for FECs. These messages are also transported over TCP. An LSR can request a label mapping from a neighboring LSR whenever it chooses (say, whenever it needs one). It can also advertise label mappings whenever it wishes an LDP peer to use a label mapping.

Notification messages are also sent over TCP and provide status, diagnostic, and error information.

FECs, LABEL SPACES, AND IDENTIFIERS

In Chapter 4, we saw that user packets are mapped to an LSP, and that the method to achieve this mapping is to specify an FEC for each LSP. Earlier chapters also explained that an FEC could be a set of addresses, port numbers, or the PID. LDP takes a more restricted view of the FEC elements and defines two: an IP address prefix and a host address.

Mapping Rules

The following rules are set forth for the mapping of a specific packet to a specific LSP. Each rule is applied in turn until the packet can be mapped to an LSP.

- If there is exactly one LSP that has a host address FEC element identical to the packet's destination address, then the packet is mapped to that LSP.
- If there are multiple LSPs, each containing a host address FEC element that is identical to the packet's destination address, then the packet is mapped to one of those LSPs. The procedure for selecting one of those LSPs is not defined by LDP.
- If a packet matches exactly one LSP, the packet is mapped to that LSP.
- If a packet matches multiple LSPs, it is mapped to the LSP whose matching prefix is the longest. If there is no one LSP whose matching prefix is longest, the packet is mapped to one from the set of LSPs whose matching prefix is longer than the others. The procedure for selecting one of those LSPs is not defined by LDP.
- If it is known that a packet must traverse a particular egress router and there is an LSP having an address prefix FEC element that is an address of that router, then the packet is mapped to that LSP. The procedure for obtaining this knowledge is not defined by LDP.

Label Spaces and Identifiers

Label spaces in LDP are the same as those defined in MPLS, namely, the per-interface label space and the per-platform label space. A label space is identified with a 6-octet LDP identifier. The first 4 octets identify an LSR and must be a globally unique value, such as an IP address (a router ID). The last two octets identify the label space within the LSR. These octets are set to zero for platform-wide label space.

If the LSR uses multiple label spaces, it associates a different LDP identifier with each label space. Multiple label spaces can be encountered in ATM networks in which two ATM switches have multiple links connecting them and perhaps reuse the labels on each interface (see Figure 5–4). With this approach, a label space and its LSR is always known if the LDP identifier accompanies an LDP message. In this example, labels 1 through 500 are used twice; the LDP identifiers keep the label spaces uniquely identified.

Figure 5–4   Multiple label spaces.

LDP SESSIONS

LSRs establish sessions between them to advertise and exchange labels. Each label space advertising and exchange operation requires a separate LDP session. As noted earlier, the LDP session runs over TCP.

Sessions Between Indirectly Connected LSRs

Chapter 4 introduced the concept of MPLS tunnels and label stacking. Figure 5–5 shows how LSRs not directly connected advertise labels. We assume LSR A and LSR D want to set up an LSP between them. LSR B and LSR C are intermediate LSRs between A and D.

LSR A applies two labels on the LSP toward LSR D, labels 33 and 21. In event 1, LSR A learns about label 21 from LSR D. In event 2,

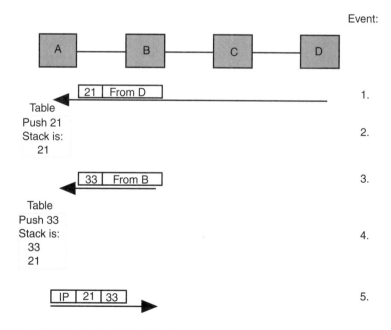

Figure 5–5   Label distribution: Adjacent and nonadjacent LSRs.

LSR A pushes 21 into its label stack. In event 3, LSA A learns about label 33 from LSR B. In event 4, LSA A pushes this label into its label stack.

When LSR A sends traffic to LSR D, it conveys labels 33 and 21 in the packet header. Label 33 is used between LSR A and LSR B, and label 21 is used between LSR A and LSR D, shown in event 5. The other labels used between B, C, and D are not shown in this example.

Node A knows which node (B or D) is involved in the label relationship because node A has established a peer adjacency relationship with nodes B and D. Of course, the fact that label 21 is to be pushed so that it is not processed by nodes B and C must be determined by the network administrator if label stacking is needed. If so, the participating LSRs must be so configured.

How LSRs Know About Other LSRs

LSRs discover each other in one of two ways. The basic discovery mechanism is used when LSR neighbors are connected directly by a link. An LSR periodically sends LDP Hello messages out of its interfaces. As noted, these Hellos are sent over UDP, with a multicast address of "all routers on this subnet." The Hello contains the LDP identifier discussed earlier.

The second method is an extended discovery mechanism. The LSR must send a Hello message, called a *targeted Hello* (as in Figure 5–5 for label 21), to LSRs with a specific IP address; the message contains the LDP identifier. These targeted addresses have been discovered by conventional routing protocols.

LDP LABEL DISTRIBUTION AND MANAGEMENT

This part of the chapter explains the LDP conventions for label distribution and continues the introduction to the subject provided in Chapter 4. The organization of the explanations is listed below in the order of discussion.

- Independent control mapping
- Ordered control mapping
- Downstream-on-demand label advertisement
- Problem management for the first three distribution methods
- Downstream unsolicited label advertisement

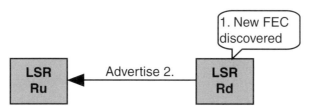

(a) Downstream unsolicited advertisement mode

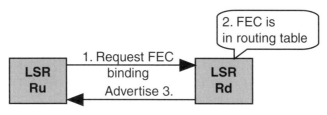

(b) Or the Ru issues a request

Figure 5–6   Independent control mapping.

Independent Control Mapping

If an LSR is configured for independent control, a mapping message is transmitted by the LSR upon any of the following conditions:

- The LSR recognizes a new FEC in the IP routing table, and the label advertisement mode is downstream unsolicited advertisement, as shown in Figure 5–6(a).
- The Rd LSR receives an LDP request message from an upstream (Ru) peer for a FEC present in the Rd LSR's routing table, as shown in Figure 5–6(b).
- The next hop for a FEC changes to another LDP peer, and loop detection is configured.
- The attributes (such as a path vector or hop count) of a mapping change.
- The receipt of a mapping from the downstream next hop and (a) no upstream mapping has been created, (b) loop detection is configured or (c) the attributes of the mapping have changed.

Ordered Control Mapping

A mapping message is transmitted by downstream LSRs upon any of the following conditions:

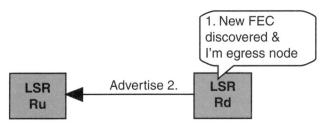

(a) Downstream unsolicited advertisement mode

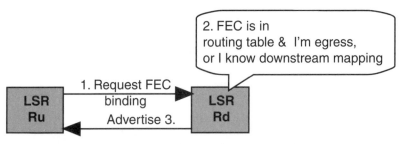

(b) Downstream on demand advertisement mode

Figure 5–7   Ordered control mapping.

- The LSR recognizes a new FEC in the routing table and is the egress for that FEC (as shown in Figure 5–7(a)).
- The LSR receives a request message from an upstream peer for an FEC present in the LSR's forwarding table, and the LSR is the egress for that FEC or has a downstream mapping for that FEC (as shown in Figure 5–7(b)).
- The next hop for a FEC changes to another LDP peer, and loop detection is configured.
- The attributes (such as a path vector or hop count) of a mapping change.
- The receipt of a mapping from the downstream next hop and (a) no upstream mapping has been created, (b) loop detection is configured or (c) the attributes of the mapping have changed.

Downstream-on-Demand Label Advertisement

This operation is straightforward; the upstream LSR is responsible for requesting label mappings, as shown in Figure 5–8. This operation is often used in ATM-based networks, and is explained in Chapter 6 (see Figure 6–5).

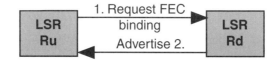

Figure 5–8   Downstream-on-demand label assignment.

Problem Management for the First Three Distribution Methods

As you can see thus far in the discussions of LDP operations, considerable flexibility is allowed in how FEC/label bindings are set up. This flexibility can create some problems.

Neighboring LSRs can get into a lockout situation such that no labels are distributed. For example, in Figure 5–9, Ru is using downstream unsolicited advertisement mode and Rd is using downstream-on-demand mode. Rd may assume that Ru will request a label mapping when it wants one, and Ru may assume that Rd will advertise a label if it wants Ru to use one. If this situation occurs, no labels will be distributed between these nodes.

The obvious solution is for the two LSRs to agree on the modes to be used between them, based on the common sense approach that an LSR configured for downstream-on-demand mode should not be expected to send unsolicited advertisements.

Downstream Unsolicited Label Advertisement

This operation was highlighted in Chapter 4 when LIBs and LFIBs were introduced. Recall that the downstream LSR is responsible for advertising a label mapping when it wants an upstream LSR to use the label, as shown in Figure 5–10. The LDP specification allows an upstream LSR to issue a mapping request if it so desires, but it might be ignored by the downstream LSR; in Chapter 4, we learned that for

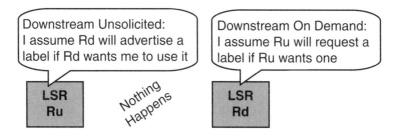

Figure 5–9   Lockout of message exchanges.

Figure 5–10   Downstream unsolicited label assignment.

practical reasons it is indeed usually ignored because the Ru is not in the path (the next hop) to the destination IP address.

LDP MESSAGES

LDP messages are defined in a media-independent format. The intent is that several messages can be combined in a single datagram to minimize the CPU processing overhead associated with input/output. Eleven messages are used by LDP. They are described later in this section. For now, let's examine the conventions for coding the messages.

The LDP Header

Each LDP message (called a *protocol data unit,* or PDU) begins with an LDP header, followed by one or more LDP messages. The header is shown in Figure 5–11. The fields in the PDU header specify the following parameters:

- Version: The version number of the protocol, currently version 1.
- PDU length: Total length of PDU in octets, excluding the version and length fields.
- LDP ID: Identifier of the label space of the sending LSR of this message. The first four octets contain an IP address assigned to the LSR: the router ID. The last two octets identify a label space

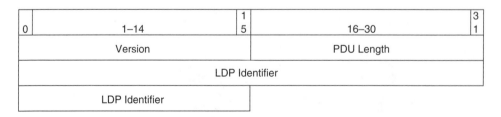

Figure 5–11   LDP header.

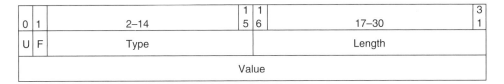

0	1	2–14	1 5	1 6	17–30	3 1
U	F	Type			Length	
			Value			

Figure 5–12 Type-length-value (TLV) encoding.

within the LSR. For a platformwide label space, these fields should both be 0.

Type-Length-Value (TLV) Encoding

LDP uses a type-length-value (TLV) encoding scheme to encode much of the information carried in LDP messages. As shown in Figure 5–12, the LDP TLV is encoded as a 2-octet field that uses 14 bits to specify a type and 2 bits to specify behavior when an LSR doesn't recognize the type, followed by a 2-octet-length field, followed by a variable-length value field.

Upon receipt of an unknown TLV, if the U bit (unknown) is set to 0, a notification must be returned to the message originator and the entire message must be ignored. If the U bit is set to 1, the unknown TLV is silently ignored and the rest of the message is processed as if the unknown TLV did not exist.

The forward unknown (F) TLV bit applies only when the U bit is set and the LDP message containing the unknown TLV is to be forwarded. If F is clear (= 0), the unknown TLV is not forwarded with the containing message; if F is set (= 1), the unknown TLV is forwarded with the containing message.

The LDP Message Format

All LDP messages have the same format, shown in Figure 5–13. The fields in the message specify the following:

- U bit: The unknown message bit. If this bit is set to 1 and it is "unknown" (message cannot be interpreted) to the receiver, then the receiver silently discards the message.
- Message type: The type of message.
- Message length: Length of message ID, mandatory parameters, and optional parameters.

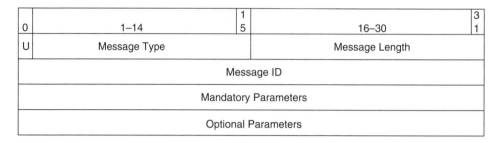

Figure 5–13   LDP message format.

- Message ID: A unique identifier of this message. The field can be used to associate Notification messages with another message.
- Mandatory parameters: Set of mandatory parameters, explained in the next section.
- Optional parameters: Set of optional parameters, explained in the next section.

In principle, everything appearing in an LDP message could be encoded as a TLV, but the LDP specification does not use the TLV scheme at all times. It is not used when it is unnecessary and its use would waste space. For example, it is not necessary to use the TLV format if the length of the value is fixed or if the type of the value is known and does not have to be assigned a type ID.

TLVs: Formats and Functions

This part of the chapter summarizes the functions of the TLVs. Recall that all TLVs use the format shown in Figure 5–12.

- **FEC:** This TLV carries the FECs that are exchanged between LSRs. Remember that MPLS and LDP use only addresses for an FEC, and not port numbers or PIDs. The FEC can be an address prefix or a full host address. It can also contain addresses pertaining to other networks, such as IPX, but there is little interest in any address other than IP. The FEC can be coded as a wildcard, in which case all FECs that previously have been associated with a label are identified. This feature is a handy tool for label withdrawal and label release operations.

- **Address list:** The address list TLV appears in Address and Address Withdraw messages. Currently, only IPv4 and IPv6 are defined for this TLV.

- **Hop count:** This TLV appears in messages that set up LSPs. It calculates the number of LSR hops along an LSP as the LSP is being set up. It can be used for loop detection, described in the "Loop Detection and Control" section.

- **PATH vector:** This TLV is also used for loop detection with the hop count TLV in the Label Request and Label Mapping messages. Its use in the Label Request message records the path of LSRs the request has traversed. Its use in the Label Mapping message records the path of the LSRs that a label advertisement has traversed to set up an LSP.

- **Generic label:** This TLV contains labels for use on links for which label values are independent of the underlying link technology (a bearer service), such as PPP and Ethernet links.

- **ATM label:** If ATM is used as a bearer service, this TLV contains ATM VPI/VCI values.

- **Frame Relay label:** If Frame Relay is used as a bearer service, this TLV contains Frame Relay DLCI values.

- **Status:** This TLV is used for diagnostic purposes, such as to report the success or failure of an operation.

- **Extended status:** This TLV extends the status TLV by providing additional bytes for more status information.

- **Returned PDU:** This TLV can operate with the status TLV. An LSR uses this parameter to return part of an LDP PDU to the LSR that sent it. The value of this TLV is the PDU header and as much PDU data following the header as is appropriate for the condition being signaled by the Notification message.

- **Returned message:** This TLV can also be used with the status TLV. An LSR uses this parameter to return part of an LDP message to the LSR that sent it.

- **Common Hello parameters:** Recall that neighbor LSRs can periodically send Hello messages to each other to make sure they are up and running. This TLV contains common parameters to manage details of this operation, such as determining how often Hellos are sent and received and recording how many have been sent and received during a specified period of time.

- **IPv6/IPv4 transport address:** If IPv6 addresses are used, this TLV allows an IPv6 address to be used when TCP is opened for an LSP session. If the TLV is not present, the source address in the IP header is used. The same idea holds for IPv4 addresses.

- **Common session parameters:** This TLV contains values proposed by the sending LSR for parameters that must be negotiated for every LDP session. These parameters are as follows.

  (a) Keep-alive time: Specifies the maximum number of seconds that can elapse between the receipt of successive PDUs from the LDP peer on the session TCP connection. The keep-alive timer is reset each time a PDU arrives.

  (b) Label advertisement discipline: Specifies downstream unsolicited, or downstream on demand.

  (c) Loop detection: Specifies whether loop detection is enabled or disabled.

  (d) PATH vector limit: Specifies the maximum path vector length.

  (e) Maximum PDU length: Specifies the maximum length of the LDP PDU.

  (f) Receiver LDP identifier: Identifies the receiver's label space.

      This LDP ID, together with the sender's LDP ID in the PDU header, enables the receiver to match the Initialization message with one of its Hello adjacencies.

- **ATM session parameters:** This TLV specifies the ATM merge capabilities of an ATM-LSR. The options are as follows.

  (a) Merge not supported

  (b) VP merge supported

  (c) VC merge supported

  (d) VP and VC merge supported

  This TLV provides information about VC directionality, which means the use of a VCI in one direction or in both directions on the link. It also contains a field that specifies the range of ATM labels supported by the sending LSR.

- **Frame Relay session parameters:** This TLV contains the same type of information as that of the ATM session parameters, except that the TLV pertains to DLCIs.

- **Label Request message ID:** The value of this parameter is the message ID of the corresponding Label Request message.

- **Private:** Vendor-private TLVs and messages that convey vendor-private information between LSRs.
- **Configuration sequence number:** This TLV specifies a configuration sequence number that identifies the configuration state of the sending LSR. It is used by the receiving LSR to detect configuration changes on the sending LSR.

The LDP Messages: Formats and Functions

This section provides information on the formats and functions of the LDP messages listed below.

- Notification
- Hello
- Initialization
- KeepAlive
- Address
- Address Withdraw
- Label Mapping
- Label Request
- Label Withdraw
- Label Release
- Label Abort Request

Notification Message. The Notification message is used by an LSR to notify its peer about unusual or error conditions. Examples of conditions are (a) receipt of unknown, erroneous, or malformed messages, (b) expiration of a keep-alive timer, (c) a shutdown by a node, and (d) failure of an LSP session initialization. In some situations, the LSR may terminate the LDP session (closing the TCP connection). The format for this message is shown in Figure 5–14.

The message ID uniquely identifies each message. It is coded in all messages and is not explained again. The status TLV indicates the status of event. The optional parameters are these TLVs: (a) extended status, (b) returned PDU, (c) returned message.

When an LSR receives a Notification message that carries a status code that indicates a fatal error, it terminates the LDP session immediately by closing the session TCP connection and discards all states

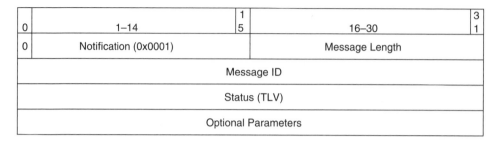

Figure 5–14   Notification message.

associated with the session, including all label-FEC bindings learned from the session.

Hello Message.   The Hello message is exchanged between two LDP peers during an LDP discovery operation. The format for this message is shown in Figure 5–15.

    The Common Hello TLV was explained earlier in this chapter. The format for this TLV is shown in Figure 5–16.

    An LSR maintains a record of Hellos received from potential LSR peers. The Hello hold time specifies the time the sending LSR will maintain its record of Hellos from the receiving LSR without receipt of another Hello. A pair of LSRs negotiates the hold times they use for Hellos from each other. Each proposes a hold time. The hold time used is the minimum of the hold times proposed in their Hellos.

    The T bit is called Targeted Hello. A value of 1 specifies that this Hello is a Targeted Hello. A value of 0 specifies that this Hello is a Link Hello. The R bit is called the R Request Send Targeted Hellos. A value of 1 requests the receiver to send periodic Targeted Hellos to the source of this Hello. A value of 0 makes no request.

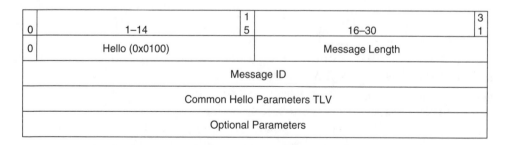

Figure 5–15   Hello message.

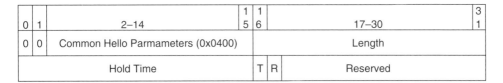

Figure 5–16   Common Hello parameters TLV.

Optional parameters are the IPv4 and IPv6 transport address TLVs and a configuration sequence number, which is used by the receiving LSR to detect configuration changes at the sending LSR.

Initialization Message.   The Initialization message is exchanged when LDP peers want to set up an LDP session. During this procedure, the LSRs negotiate parameters, such as a keep-alive timer and the types of advertisements that will be supported (downstream unsolicited or downstream on demand). If Frame Relay or ATM labels are to be used during the session, the rules for using these labels are also negotiated. The format for this message is shown in Figure 5–17.

The common session parameters TLV was explained in the previous section. Optional parameters are ATM and Frame Relay session parameters.

KeepAlive Message.   The KeepAlive message is exchanged between peers to monitor the integrity of the TCP connection supporting the LDP session. The format for this message is shown in Figure 5–18. There are no optional parameters for this message.

Address Message.   The Address message is sent by an LSR to its LDP peer to advertise its interface addresses. An LSR that receives an

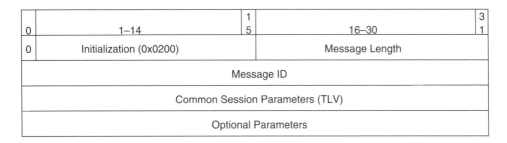

Figure 5–17   Initialization message.

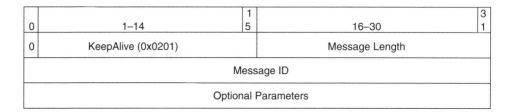

Figure 5–18   KeepAlive message.

Address message uses the addresses it learns to maintain a database for mapping between peer LDP identifiers and next hop addresses. The format for the message is shown in Figure 5–19.

The address list TLV is the list of interface IP addresses being advertised by the sending LSR. There are no optional parameters for this message.

Address Withdraw Message.   The Address Withdraw message "undoes" the Address message and withdraws a previously advertised interface address or addresses. The format for this message is shown in Figure 5–20.

The address list TLV contains the list of addresses that are being withdrawn by the sending LSR of this message. It has no optional parameters.

Label Mapping Message.   The Label Mapping message advertises to an LDP peer the FEC-label bindings to the peer. If a LSR distributes a mapping for an FEC to multiple LDP peers, it is a local matter whether the LSR maps a single label to the FEC and distributes that mapping to all its peers or whether it uses a different mapping for each of its peers.

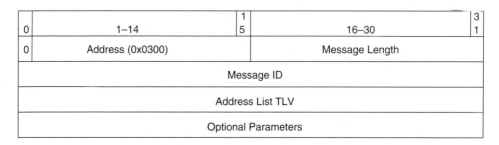

Figure 5–19   Address message.

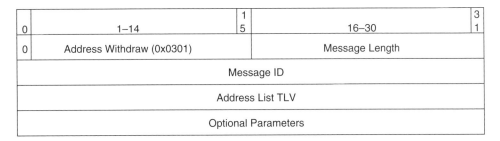

Figure 5–20   Address Withdraw message.

An LSR receiving a Label Mapping message from a downstream LSR for a prefix or host address FEC element should not use the label for forwarding unless its routing table contains an entry that exactly matches the FEC value. The format for the Label Mapping message is shown in Figure 5–21.

Of course, this message must contain the IP addresses and their associated labels. The FEC TLV specifies the FEC part of the FEC-label mapping being advertised. The Label TLV specifies the label part of the FEC-label mapping. The optional TLVs are the Label Request message ID, the hop count, and the path vector.

Label Request Message.   The Label Request message is used by an LSR to request that an LDP peer furnish a label binding for an FEC. An LSR may transmit a Request message under any of the following conditions:

- The LSR recognizes a new FEC through the forwarding table, the next hop is an LDP peer, and the LSR does not have a mapping from the next hop for the given FEC.

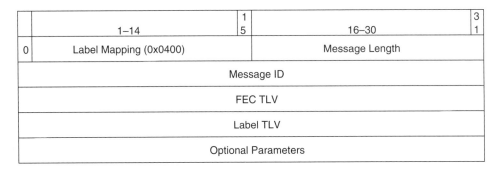

Figure 5–21   Label Mapping message.

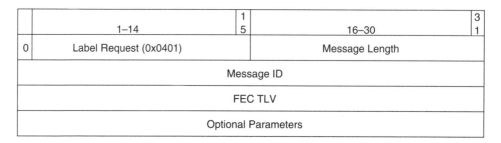

Figure 5–22   Label Request message.

- The next hop to the FEC changes, and the LSR does not have a mapping from that next hop for the given FEC.
- The LSR receives a Label Request for an FEC from an upstream LDP peer, the FEC next hop is an LDP peer, and the LSR does not have a mapping from the next hop.

The format for the Label Request message is shown in Figure 5–22.

The FEC TLV identifies the label value being requested. The optional TLVs are hop count and path vector.

Label Withdraw Message.   The Label Withdraw message destroys a mapping between FECs and labels. It is sent to an LDP peer to inform that node not to continue to use specific FEC-label bindings that the LSR had previously advertised. An LSR transmits a Label Withdraw message under the following conditions:

- The LSR no longer recognizes a previously known FEC for which it has advertised a label.
- The LSR has decided unilaterally (e.g., by configuration) to no longer label-switch an FEC (or FECs) with the label mapping being withdrawn.

The FEC TLV specifies the FEC for which labels are to be withdrawn. If no label TLV follows the FEC, all labels associated with the FEC are to be withdrawn; otherwise, only the label specified in the optional label TLV is to be withdrawn. An LSR that receives a Label Withdraw message must respond with a Label Release message. The format for the Label Withdraw message is shown in Figure 5–23. The fields and TLVs in the message are explained in previous text.

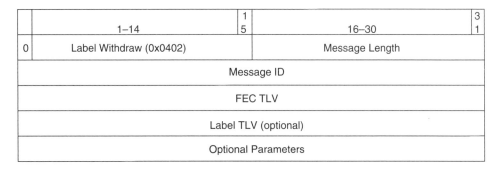

Figure 5–23   Label Withdraw message.

Label Release Message.   The Label Release message, depicted in Figure 5–24, informs the receiving LDP peer that the LSR no longer needs specific FEC-label bindings. An LSR must transmit a Label Release message under any of the following conditions:

- The LSR that sent the label mapping is no longer the next hop for the mapped FEC, and the LSR is configured for conservative operation.
- The LSR receives a label mapping from an LSR that is not the next hop for the FEC and the LSR is configured for conservative operation.
- The LSR receives a Label Withdraw message.

The TLVs of this message are described in previous text. The optional parameter is the label TLV.

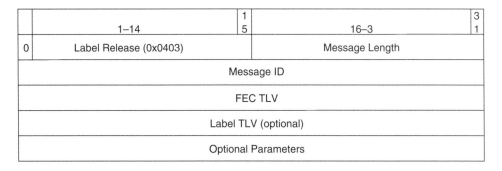

Figure 5–24   Label Release message.

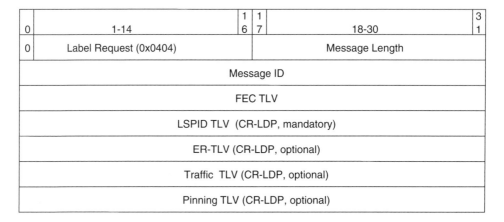

0	1-14	1 6	1 7	18-30	3 1
0	Label Request (0x0404)			Message Length	
Message ID					
FEC TLV					
LSPID TLV (CR-LDP, mandatory)					
ER-TLV (CR-LDP, optional)					
Traffic TLV (CR-LDP, optional)					
Pinning TLV (CR-LDP, optional)					

Figure 5–25   Label Abort Request message.

Label Abort Request Message.   This message aborts an out-standing Label Request message. There are a variety of reasons for issuing an abort message, such as an OSPF or BGP prefix advertisement that changes the label request operation. The format for this message is shown in Figure 5–25. The contents of the message are explained in previous text and in the traffic engineering discussion in this book (Chapter 7), and there are no optional parameters.

LOOP DETECTION AND CONTROL

Traffic looping is potentially a very big problem in internets. The phenomenon is not uncommon during periods of route changes and the associated time to stabilize the routing tables to reach convergence. Looping can also occur if routing protocols or routing tables are misconfigured. Looping means traffic traverses through the network endlessly, going, say, from node A, to B, to C, and back to A, then B, and so on.

Several mechanisms have been developed to prevent excessive looping, and LDP supports two of them. They are listed below and explained in this part of the chapter:

- Buffer control
- Time to live (TTL), supported by LDP
- Path vectors, supported by LDP
- Colored threads

Buffer Control

Buffer control is employed in some ATM networks because ATM was not designed to detect and prevent loops. This approach limits the amount of buffer space allotted to packets, which results in excess packets being discarded. Since the LSR is not processing as many packets, it can devote more of its resources to stabilizing the LIB and LFIB and achieving network convergence.

Time to Live (TTL)

LDP can control looping with a time to live field. TTL works in label switching networks just as it does in IP forwarding networks, except the TTL value is called a hop count and is incremented instead of decremented. The originator of the traffic sets a value in the hop count in the packet. When an LSR propagates a message containing a hop count field, it increments the count. If an LSR detects a hop count value that has reached a configured maximum value, the packet is discarded.

The value of the TTL field is usually a default parameter that is established by the MPLS/LDP software designer or by the network operator. A hop count of 0 is interpreted to mean the hop count is unknown. See the sidebar titled "Hop Count Ideas" for more information on this topic.

Path Vectors

If you are familiar with the well-known concept of route recording, used in some internets and SS7 networks, path vectors will be familiar territory. A field in the LSP message called the path vector contains a list (a unique ID) of the LSRs that the packet has traversed, as shown in Figure 5–26(a). When an LSR propagates a message containing a path vector field, it adds its LSR ID to the list. An LSR that receives a message with a path vector that contains its LSR ID detects that the packet has traversed a loop.

In Figure 5–26(a), the first five events have resulted in each LSR's ID being inserted into the path vector list and the packet being sent to a next node. The packet is in a loop, because in event 6, it is returned to node D. This node has already inserted its ID into the path vector list in event 2, so it knows the packet is in a loop.

Colored Threads

Colored threads are similar to path vectors in that they depend on an LSR detecting a loop by examining a value in the packet that the LSR knows it has received once before and so it detects that the packet has

---

Hop Count Ideas (section 3.4.4.1 of RFC 3036)

The first LSR in the LSP (ingress for a Label Request message, egress for a Label Mapping message) should set the hop count value to 1.

By convention a value of 0 indicates an unknown hop count. The result of incrementing an unknown hop count is itself an unknown hop count (0).

Use of the unknown hop count value greatly reduces the signaling overhead when independent control is used. When a new LSP is established, each LSR starts with unknown hop count. Addition of a new LSR whose hop count is also unknown does not cause a hop count update to be propagated upstream since the hop count remains unknown. When the egress is finally added to the LSP, then the LSRs propagate hop count updates upstream via Label Mapping messages.

Without use of the unknown hop count, each time a new LSR is added to the LSP a hop count update would need to be propagated upstream if the new LSR is closer to the egress than any of the other LSRs. These updates are useless overhead since they don't reflect the hop count to the egress.

From the perspective of the ingress node, the fact that the hop count is unknown implies nothing about whether a packet sent on the LSP will actually make it to the egress. All it implies is that the hop count update from the egress has not yet reached the ingress.

If an LSR receives a message containing a Hop Count TLV, it must check the hop count value to determine whether the hop count has exceeded its configured maximum allowable value. If so, it must behave as if the containing message has traversed a loop by sending a Notification message signaling Loop Detected in reply to the sender of the message.

---

traversed a loop. The value is called a colored thread; it is simply (a) the IP address of the originating LSR and (b) a unique, unambiguous value that the LSR assigned to the packet.

Like path vectors the color thread value is used to detect a loop, as shown in Figure 5–26(b). The first five events have resulted in each LSR (a) examining the colored thread, (b) remembering it, and (c) perhaps sending the packet to a next node. Obviously, the packet is in a loop, because in event 6, it is returned to node D. This node has stored the fact that it has once received this colored thread, so it knows the packet is in a loop.

At first glance, it is difficult to tell the difference between path vectors and colored threads. Upon closer examination, it can be seen that colored threads do not consume as much overhead as do path vectors. The colored threads overhead field is just the originating LSR ID and a unique

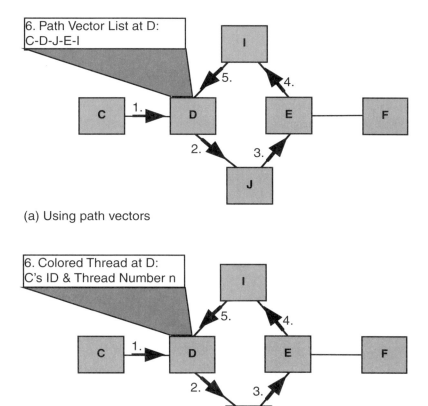

(a) Using path vectors

(B) Using colored threads

Figure 5–26   Path vectors and colored threads.

number. The path vector contains the ID of every LSR on the path, so it does not scale as well in a large routing domain with many LSRs.

Colored threads have not yet been added to the LDP RFC, but it is a simple and effective technique that will likely find its way into the Internet standards.

RSVP AND LABEL DISTRIBUTION

As its name implies, the Resource Reservation Protocol reserves resources for a session (flow) in an Internet. This aspect of the Internet is quite different from the underlying design intent of the system, which as we learned earlier, was established to support only a best-effort service, without regard to predefined requirements for the user application.

RSVP is intended to guarantee performance by reserving the necessary resources at each machine that participates in supporting the flow of traffic (such as a video or audio conference). Remember that IP is a connectionless protocol that does not set up paths for the traffic flow, whereas RSVP establishes these paths and guarantees the bandwidth on the paths.

RSVP does not provide routing operations but uses IPv4 or IPv6 as the transport mechanism in the same fashion as do the Internet Control Message Protocol (ICMP) and the Internet Group Message Protocol (IGMP).

RSVP requires the receivers of the traffic to request QOS for the flow. The receiver host application must determine the QOS profile, which is then passed to RSVP. After the analysis of the request, RSVP sends request messages to all the nodes that participate in the data flow.

RSVP operates with unicast or multicast procedures and interworks with current multicast protocols. RSVP requires the receivers of the traffic to request QOS for the flow. The receiver host application must determine the QOS profile, which is then passed to RSVP. After the analysis of the request, RSVP sends request messages to all the nodes that participate in the data flow.

Aspects of RSVP Pertinent to MPLS

This section provides a brief and general overview of those RSVP features that are pertinent to using RSVP with MPLS. The reader who is experienced with RSVP can skip to the next section.

RSVP is a signaling protocol used to set up QOS reservations in an internet. As shown in Figure 5–27, quality of service is implemented for a particular data flow by mechanisms collectively called "traffic control." These mechanisms include (1) a classifier, (2) admission control, (3) a packet scheduler (or some other link-layer-dependent mechanism to determine when particular packets are forwarded), and (4) policy control.

The classifier determines the QOS class (and perhaps the route) for each packet, based on examination of the IP and transport layer headers. For each outgoing interface, the packet scheduler or other link-layer-dependent mechanism achieves the promised QOS. The packet

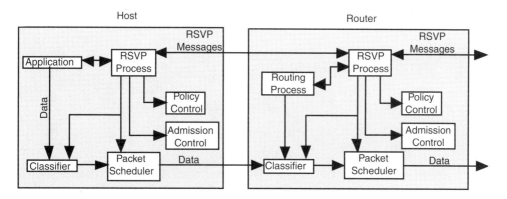

Figure 5–27   RSVP operational entities.

scheduler implements QOS service models defined by the Integrated Services Working Group.

During reservation setup, an RSVP QOS request is passed to two local decision modules: admission control and policy control. Admission control determines whether the node has sufficient available resources to supply the requested QOS. Policy control determines if a flow is permitted according to administrative rules, such as certain IP addresses are or are not permitted to reserve bandwidth; certain protocol IDs are or are not permitted to reserve bandwidth, and so on.

Sessions.   RSVP defines a session to be a flow with a particular IP address destination and transport layer protocol. An RSVP session is defined by IP destination address (DestAddress), IP protocol ID (ProtocolId), and destination port ID (DstPort). The IP destination address of the data packets can be a unicast or multicast address. The ProtocolId is the IP protocol ID. The optional DstPort parameter is a "generalized destination port." DstPort could be defined by a UDP/TCP destination port field, by an equivalent field in another transport protocol, or by some application-specific information.

The Key RSVP Messages.   RSVP requires the receivers of the traffic to request QOS for the flow. The receiver host application must determine the QOS profile, which is then passed to RSVP. After the analysis of the request, RSVP sends request messages to all the nodes that participate in the data flow. As Figure 5–28 shows, the operations begin with the RSVP Path message. It is used by a server (the flow sender) to set up a path for the session.

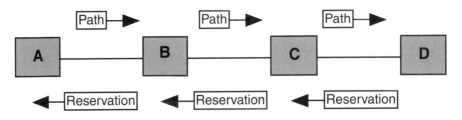

Figure 5–28   Path and reservation messages.

Figure 5–28 also shows that the Reservation messages are sent by the receivers of the flow, and they allow sender and intermediate machines (such as routers) to learn the receivers' requirements. The route taken back to the server in the network by the Reservation message is the same route taken by the Path message.

Admission Control and Policy Control. Looking at Figure 5–27, we see that the RSVP process passes the request to admission control and policy control. If either test fails, the reservation is rejected and the RSVP process returns an error message to the appropriate receiver(s). If both succeed, the node sets the packet classifier to select the data packets defined by the *filterspec* and interacts with the appropriate link layer to obtain the desired QOS defined by the *flowspec*. Let's examine the filterspec and the flowspec.

The Flow Descriptor

A simple RSVP reservation request consists of a flowspec together with a filterspec; this pair is called a *flow descriptor*. See Figure 5–29. The flowspec specifies a desired QOS. The filterspec, together with a session specification, defines the set of data packets—the flow—to receive the QOS defined by the flowspec.

The flowspec sets parameters in the node's packet scheduler or other link layer mechanism, and the filterspec sets parameters in the packet classifier. Data packets that are addressed to a particular session but do not match any of the filterspecs for that session are handled as best-effort traffic.

The flowspec in a reservation request will generally include a service class and two sets of numeric parameters: (1) an *Rspec* (R for "reserve") that defines the desired QOS and (2) a *Tspec* (T for "traffic") that describes the data flow. The formats and contents of Tspecs and

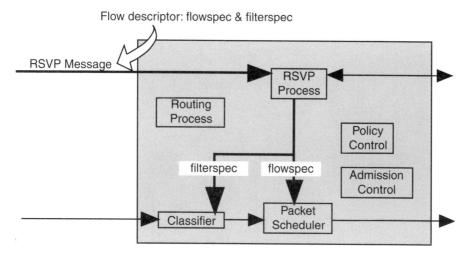

Figure 5–29   Flow descriptor.

Rspecs are determined by the integrated service models (see RFC 2210).

The fields in the RSVP messages are called objects. Since the publication of RSVP (see RFC 2205), many objects have been added to the original specification. Those relating to MPLS are explained in this discussion.

Interworking Concepts for MPLS and RSVP

The extensions to RSVP to support MPLS label switching set up LSPs with or without resource reservations. The extensions also provide for LSP rerouting, load balancing, constrained routing (dictating the route of the LSP in the routing domain), and loop detection. These extensions to RSVP mirror many of the operations in the LDP explained in the first part of this chapter.

[AWDU01] provides the rules on the interworking of RSVP and MPLS. In addition, several other IETF working drafts define other rules. This part of the chapter highlights [AWDU01] and cites other references where appropriate.

Hosts and routers that support both RSVP and MPLS can associate labels with RSVP flows. Once an LSP is established, the traffic through the path is defined by the label applied at the ingress node of the LSP. The set of packets that are assigned the same label value belong to the same FEC and are the same as the set for the RSVP flow. When labels are associated with traffic flows, it is possible for a router to identify the appropriate RSVP reservation state for a packet, based on the packet's label value.

The RSVP/MPLS model uses downstream-on-demand label distribution. Referring to Figure 5–28, we see that the upstream nodes ask for (demand) a label binding (A to B, B to C, etc.). A request to bind labels to a specific LSP tunnel is initiated by an ingress node (in Figure 5–28, node A), through the RSVP Path message, which contains a LABEL_ REQUEST object. This object contains suggested label values, including ATM and Frame Relay virtual circuit numbers (if needed).

Labels are allocated downstream and distributed (propagated upstream) by the Reservation message. For this purpose, the RSVP Reservation message is extended with a LABEL object. This object contains the label that is to be used between neighbor nodes. For example, in Figure 5–28, the Path message between nodes B and C contains the LABEL_REQUEST object, and the Reservation message contains the LABEL object.

The LABEL object is inserted into the filterspec list immediately following the filterspec to which it pertains. The receipt of the label allows the node to update its ILM (for a review of the ILM refer to Chapter 4, Figure 4–16).

Explicit Routing. This extension of RSVP also supports explicit routing, known as constrained routing in MPLS jargon. This operation is executed by placement of the EXPLICIT_ROUTE object into Path messages, as shown in Figure 5–30 where nodes D, J, E, and F are established for the LSP.

The EXPLICIT_ROUTE object contains the hops for the explicitly routed LSP. The explicitly routed path can be administratively specified or automatically computed by a suitable entity based on QOS and policy requirements, taking into consideration the prevailing network state, but RSVP does not define how the explicit path is determined. However, the hops for the explicit route are identifed as (a) an IPv4 prefix, (b) and IPv6 prefix, or (c) an Autonomous System number.

In addition, explicit routing permits the use of loose or strict routing. Its function is similar to IP's (seldom used) source routing option. Loose routing is a suggested set of hops, and strict routing is a required set of hops.

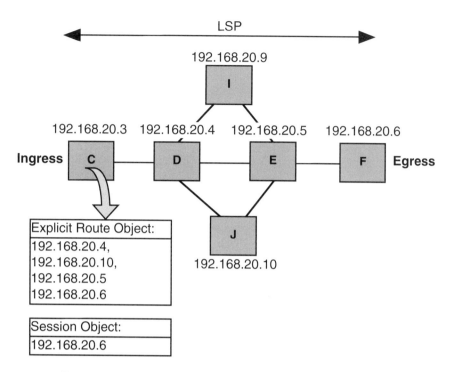

Figure 5–30   RSVP EXPLICIT_ROUTE and SESSION.

Specifying Ingress and Egress Nodes.   The SESSION object, also shown in Figure 5–30, is a useful field for the network administrator who wants to control the ingress and egress nodes for the LSP but not necessarily control each node between the ingress-egress pairs. To execute this option, the SESSION object contains the IP address of the egress node.

Priorities for Session.   Another field defined for this RSVP extension is SESSION_ATTRIBUTE. It is used by the RSVP/MPLS nodes to identify a priority for the LSP/flow with respect to consuming resources at the nodes. It is used to determine whether the session can preempt another session.

More Information About RSVP Extensions

It is time to leave the discussion of RSVP extensions for MPLS networks, as least for the time being. The subject is examined again in Chapter 7 with regard to traffic engineering and the use of other RSVP extensions for setting up and using backup LSPs in the routing domain (see the "Using RSVP to Establish Alternate/Detour LSPs" section in Chapter 7).

BGP AND LABEL DISTRIBUTION

The Border Gateway Protocol has also been enhanced to support label distribution [REKH00]. This part of the chapter provides a summary of the main points of this working draft, and Chapter 12 provides more information on other related drafts and RFCs. If you need background information on BGP, take a look at one of my companion books in this series, *IP Routing Protocols*.

When BGP is used to distribute a particular route, it can also be used to distribute an MPLS label that is mapped to that route. The label mapping information for a particular route is piggybacked in the same BGP Update message that distributes the route itself.

The BGP operations are quite similar to the conventional MPLS label stacking operations. For example, if exterior router A needs to send a packet to destination D and if A's BGP next hop for D is exterior router B and if B has mapped label L to D, then A first pushes L onto the packet's label stack. A then consults its IGP to find the next hop to B—call it C. If C has distributed to A an MPLS label for the route to B, A can push this label on the packet's label stack and then send the packet to C.

If a set of BGP speakers are exchanging routes through a route reflector, then if the label distribution is piggybacked on the route distribution, the route reflector can distribute the labels as well. This improves scalability significantly.

Label distribution can be piggybacked in the BGP Update message by means of the BGP-4 Multiprotocol Extensions attribute (see RFC 2283). The label is encoded into the NLRI field of the attribute, and the SAFI (subsequent address family identifier) field indicates that the NLRI contains a label. A BGP speaker may not use BGP to send labels to a particular BGP peer unless that peer indicates, through BGP capability negotiation, that it can process Update messages with the specified SAFI field.

The Cisco router uses some of the concepts just described. For routes learned with BGP, the Cisco router does not assign a label. The ingress LSR simply uses the label assigned in the BGP next hop to label the packets that are forwarded to the BGP destination.

BGP plays a major role in virtual private networks (VPNs). This topic is covered in Chapter 12.

SUMMARY

This chapter examined three methods for label distribution: the Label Distribution Protocol (LDP), the Resource Reservation Protocol (RSVP), and the Border Gateway Protocol (BGP). The bulk of the material was devoted to LDP, because of its many messages and rules.

6

MPLS and ATM and Frame Relay Networks

Since ATM is widely used in wide area networks and performs the job of forwarding traffic (cell switching) through a network, an important aspect of MPLS is its ability to operate over ATM networks, a concept called *overlay*. It is a good idea to be able to integrate MPLS and ATM operations in one switch, instead of running MPLS on a router and ATM in a backbone cell switch. This integration is not an insurmountable problem, but it is not a trivial task. This chapter explains how these operations occur and how ATM and MPLS can interwork with each other in one node or between nodes. These nodes are called ATM-LSRs. Examples for using MPLS with Frame Relay are also provided.

Several Internet drafts and RFCs have been published to provide guidance on the interworking of MPLS with ATM and Frame Relay networks. This chapter provides a summary and tutorial on these specifications. The explanations of Frame Relay follow those on ATM and are more terse, since many of the Frame Relay operations are almost identical to ATM. In addition, Appendix B provides yet more details on the relationships of MPLS to ATM and Frame Relay.

For more details on this subject, I refer you to these references: [NAGA01], [WIDJ99], and [DAVI99].

ASPECTS OF ATM OF INTEREST TO MPLS

This section provides an overview of those aspects of ATM that are relevant to MPLS. These topics are of interest:

- Virtual circuits, the logical connections in the network
- VPIs and VCIs, the ATM labels (virtual circuit IDs)
- The ATM cell header containing the labels
- Permanent virtual circuits (PVCs) and switched virtual calls (SVCs)

Virtual Circuits

One task of ATM is to set up a virtual circuit across the network (or networks) between the user machines. In so doing, ATM logically concatenates the physical circuits between the users into the virtual circuit. Each physical circuit that is part of the virtual circuit is called a *virtual circuit segment*. This idea is shown in Figure 6–1.

VPIs and VCIs

An ATM connection is identified through two labels, called the virtual path identifier (VPI) and virtual channel identifier (VCI). In each direction, at a given interface, different virtual paths are multiplexed by

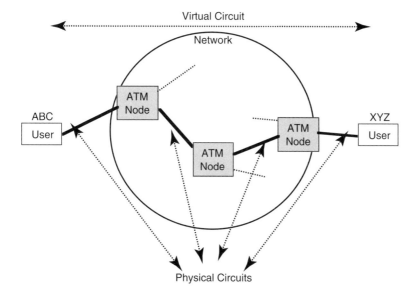

Figure 6–1   ATM virtual circuits.

ATM onto a physical circuit. The VPIs and VCIs identify these multi-plexed connections.

Virtual channel connections have end-to-end significance between two end users, usually between two ATM adaptation layer (AAL) enti-ties. The values of these connection identifiers can change as the traffic is relayed through the ATM network. For example, the specific VCI value has no end-to-end significance. It is the responsibility of the ATM net-work to keep track of the different VCI values as they relate to each other on an end-to-end basis. A good way to view the relationship of VCIs and VPIs is to think that VCIs are part of VPIs; they exist within the VPIs.

Routing in the ATM network is performed by the ATM switch exam-ining both the VCI and VPI fields in the cell or examining only the VPI field. This choice depends on how the switch is designed and whether VCIs are terminated within the network.

The VCI/VPI fields can be used with switched or nonswitched ATM operations. They can be used with point-to-point or point-to-multipoint operations. They can be preestablished (PVCs) or set up on demand, in accordance with signaling procedures, such as the B-ISDN network layer protocol (Q.2931).

Additionally, the value assigned to the VCI at the user-network interface (UNI) can be assigned by (a) the network, (b) the user, or (c) through a negotiation process between the network and the user.

To review briefly, the ATM layer has two multiplexing hierarchies: the virtual channel and the virtual path. See Figure 6–2. The VPI is a bundle of virtual channels. Each bundle must have the same endpoints. The purpose of the VPI is to identify a group of virtual channel (VC) con-nections. This approach allows VCIs to be "nailed up" end-to-end to pro-vide semipermanent connections for the support of a large number of user sessions. VPIs and VCIs can also be established on demand.

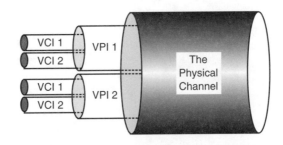

Figure 6–2   ATM connection identifiers.

The VC identifies a unidirectional facility for the transfer of the ATM traffic. The VCI is assigned at the time a VC session is activated in the ATM network. Routing might occur in an ATM network at the VC level, or VCs can be mapped through the network without further translation. If VCIs are used in the network, the ATM switch must translate the incoming VCI values into outgoing VCI values on the outgoing VC links. The VC links must be concatenated to form a full virtual channel connection (VCC). The VCCs are used for user-to-user, user-to-network, or network transfer of traffic.

The VPI identifies a group of VC links that share the same virtual path connection (VPC). The VPI value is assigned each time the VP is switched in the ATM network. Like the VC, the VP is unidirectional for the transfer of traffic between two contiguous ATM entities.

Referring to Figure 6–2, two different VCs that belong to different VPs at a particular interface are allowed to have the same VCI value (VCI 1, VCI 2). Consequently, the concatenation of VCI and VPI is necessary to uniquely identify a virtual connection.

The ATM Cell Header

The ATM PDU is called a cell; see Figure 6–3. It is 53 octets in length, with 5 octets devoted to the ATM cell header and 48 octets used by AAL and the user payload. As shown in this figure, the ATM cell is configured slightly differently for the user-network interface (the interface between the user node and the ATM switch) than for the network-to-node interface (NNI, the interface between ATM switches). Since flow control and operations, administration, and maintenance (OAM) operate at the UNI interface, a flow control field is defined for the traffic traversing this interface, but not at the NNI. The flow control field is called the generic

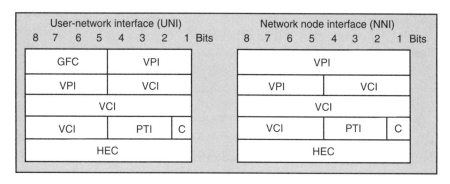

Figure 6–3   ATM protocol data units (cells).

flow control (GFC) field. If the GFC is not used, this 4-bit field is set to zeros.

Most of the values in the 5-octet cell header consist of the virtual circuit labels of VPI and VCI. Most of the VPI and VCI overhead values are available to use as the network administrator chooses. Here are some examples of how they can be used. Also refer to the sidebar titled "ATM Cannot Stack VPIs/VCIs."

Multiple VCs can be associated with one VP. This technique can be used to assign a certain amount of bandwidth to a VP and then allocate it among the associated VCs. *Bundling* VCs in VPs allows transmission of one OAM message that provides information about multiple VCs, effected by the VPI value in the header. Some implementations, to avoid processing all the bits in the VP and VC fields, do not use all the bits of VPI/VCI. Some implementations examine only the VPI bits at intermediate nodes in the network.

ATM Cannot Stack VPIs/VCIs

While on the subject of VPIs/VCIs, I should point out that one of the big differences between ATM's VPI/VCIs and the labels of MPLS is that ATM cannot support label stacking with label hierarchies. As noted in Chapter 4, MPLS label stacking is a powerful feature for supporting scaling operations, VPNs, and other services. Shortly, this subject is examined in relation to interworking of ATM and MPLS.

A payload type identifier (PTI) field identifies the type of traffic residing in the cell. The cell can contain user traffic or management/control traffic. The standards bodies have expanded the use of this field to identify other payload types (OAM, control, etc.). Interestingly, the GFC field does not contain the congestion notification codes, because the name of the field was created before all of its functions were identified. The flow control fields (actually, congestion notification bits) are contained in the PTI field.

The cell loss priority (C) field is a 1-bit value. If C is set to 1, the cell has a better chance of being discarded by the network. Whether or not the cell is discarded depends on network conditions and the policy of the network administrator. The field C set to 0 indicates a higher priority of the cell to the network.

The header error control (HEC) field is an error check field, which can also correct a 1-bit error. It is calculated on the 5-octet ATM header, and not on the 48-octet user payload. ATM employs an adaptive error detection/correction mechanism with the HEC. The transmitter calculates the HEC value on the first 4 octets of the header.

Permanent Virtual Circuits and Switched Virtual Calls

A virtual circuit can be provisioned continuously. With this approach, the user has the service of the network at any time. This concept is called a permanent virtual circuit (PVC).

A PVC is established by creation in the network nodes of entries that identify the user. These entries contain a unique identifier for the user; this identifier is known by various names such as logical channel, virtual circuit identifier (VCI), and virtual path identifier (VPI). A user need only submit this identifier to the network node. The network node examines a logical channel or virtual circuit table to discern what kind of services the user wants and with whom the user wishes to communicate.

In contrast to a PVC, a switched virtual circuit (SVC) is not preprovisioned. A user wanting to obtain network services to communicate with another user must submit a connection request message to the network. This message usually identifies the originator. It must identify the receiver, and it may also contain the virtual circuit that is to be used during the session. This virtual circuit ID value is simply a label that is used during this communications process. Once the session is over, this value is made available to any other user that wants to "pick it out" of a table. Many networks support another virtual circuit service, which is called by various names. I will use the name *semi-PVC*. With this approach, a user is preprovisioned in that the user is identified to the network as well as to the user's end communicating party. Also identified are the network features that are to be used during this session. Therefore, the network node contains information about the communicating parties and the type of services desired.

Figure 6–4 shows how the virtual circuit is initiated by the initial connection request message for an SVC. The signaling protocol used to set up the VC is Q.2931. The contents of the message are used at the ATM switches (nodes) to set up a connection and to create routing table entries. Among the fields in the Q.2931 connection request message is a destination address (called party). This address is used by the ATM node to determine the route that is to be established for the connection. Each node accepts the Q.2931 message, examines the destination address, and then consults a routing table to determine the next node that should receive the message.

Other Important Fields in the ATM Cell

ATM provides a number of attractive features that can be highlighted by reference to the connection request message in Figure 6–4. The calling party is the source address. It is used by the network to de-

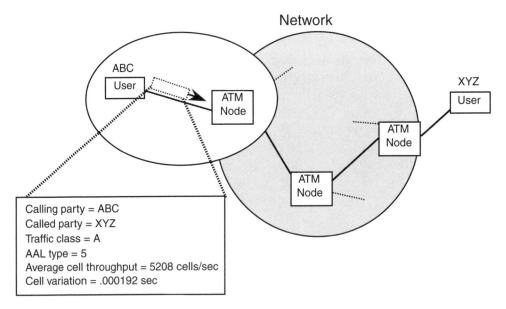

Figure 6–4   Providing information for the connection.

termine who is requesting the connection. The traffic class field in the message identifies the kind of traffic that will be sent to/from users ABC and XYZ if the connection is granted. Class A traffic is an example; it identifies synchronous, constant bit-rate traffic such as high-quality voice or video applications.

The AAL type informs the network how the user will encapsulate and format the 48-byte payload in the 53-byte cell.

The next two fields describe the quality of service needed for this connection. For this example, these two fields work hand-in-hand. The first field, titled average cell throughput, informs the network that the user wants an average cell throughput rate of 5,208 cells per second through the network. The second field, cell delay variation, places a more stringent requirement on the connection. It states that the user's application needs a variation of successive cell transfers of 192 μsec, which is no variation at all. Since .000192 × 5,208 = 1 sec, the user is defining the exact performance parameters for this class A traffic.

As each node sets up the calls, it reserves a VPI/VCI for each connection. Given an input VPI/VCI value, a node selects an unused VPI/VCI value for the output port. The next node receives this value, selects the route, and chooses the VPI/VCI values for its output port, and so on, to the final terminating user device, as shown in Figure 6–5.

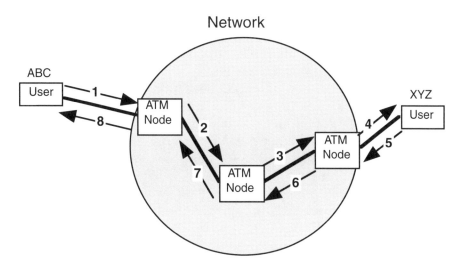

Figure 6–5   Assigning virtual circuit identifiers.

It is the job of the network to select values that are not being used on the same physical interface. This approach allows the VPI/VCI values to be reused.

ATM and MPLS: Similarities and Differences

Figure 6–5 reveals some aspects of ATM that are both different from and at the same time similar to MPLS. The arrows in events 1–4 depict the virtual circuit setup request, and the arrows in events 5–8 depict the confirmation of the virtual circuit setup. Event 1 occurs when a user specifically requests a connection.

Compare this figure with Figure 4–5 in Chapter 4, which shows that a user does not request a connection (an LSP in MPLS terms). An LSP is created if IP routing tables reveal an FEC that has next hop relationships, eventually to the destination IP address.

MPLS and ATM End-to-End Operations.   MPLS is oriented toward the behavior of an MPLS node in relation to its immediate neighbors. An MPLS node assumes, based on IP routing, that an end-to-end MPLS path will be established. If the LSP is not established for whatever reason, the MPLS node can simply resort to conventional unlabeled IP-based routing.

The ATM node is also concerned with its immediate neighbors, in that VPIs/VCIs are set up between these neighbors. However, the sequence of events in Figure 6–5 shows an ordered end-to-end dependence of all nodes. As an example, event 1 and event 8 have local significance between the two ATM nodes, but they cannot occur unless events 2–7 also take place among the other nodes.

This interdependence of nodes in an MPLS network does not exist if the network uses unsolicited, independent label assignments. If an LSR in the MPLS domain issues a binding advertisement message, it may or may not be accepted by other LSRs, depending on the relationship of the issuing LSR as a possible next hop in the path to the destination address.

Some critics of ATM have cited the end-to-end operations required to build a virtual circuit as a big drawback to the use of ATM, especially if the loss of a link or node in the end-to-end virtual circuit requires all nodes to tear down their part of the virtual circuit and start over. I do not see it that way. A well-conceived ATM network can display the same resilience as an MPLS network. Backup virtual circuits can be preconfigured or established on-the-fly, just as in MPLS. And of course, an LSP has to be established end-to-end as well.

Nonetheless, the overall features of MPLS (if used in conjunction with RSVP-TE, DiffServ, and OSPF extensions) are superior to those of ATM. I embellish on this statement in the "Principal Traffic Management Issues in and Between ATM-LSRs" section of this chapter.

Upstream Ordered VPI/VCI Allocation. In contrast to the (usual) MPLS practice of independent, downstream allocation, ATM uses upstream, ordered control allocation. The notations of events 1–8 in Figure 6–5 mean the upstream ATM node is (a) assigning a label or (b) requesting a label assignment from the downstream node. In either case, this procedure is what MPLS calls downstream on demand: a label is assigned at the instigation of the upstream node to the downstream node.

After the connection is established, the destination address is not needed in the network; only the VPI/VCI values are needed. This idea is illustrated in Figure 6–6 and should look very familiar, since it is quite similar to the MPLS operations described in earlier chapters. ATM switches use the input port and the incoming VPI/VCI value as the index into a cross-connect table, from which they obtain an output port and an outgoing VPI/VCI value. Therefore, if one or more labels can be encoded directly into the fields that are accessed by these legacy switches, then the legacy switches can, with suitable software upgrades, be used as LSRs. As noted, we refer to such devices as *ATM-LSRs*.

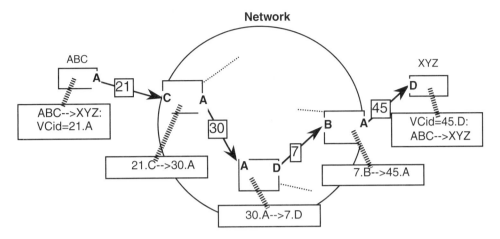

Figure 6–6   Creation of cross-connect tables.

A routing table (also called a switching table, cross-connect table, etc.) is stored at each node and reflects the state of the connection, the available bandwidth at each node, and so on. It is updated periodically as conditions in the network change. Consequently, when the Q.2931 message is received by an ATM node, it knows the "best route" for this connection—at least to the next neighbor node. These operations depend on how a vendor chooses to perform route discovery and maintain routing tables.

SCALING IP/ATM OVERLAY NETWORKS

Recall that the term *overlay* refers to running IP over (actually, through) an ATM network, as shown in Figure 6–7. The important idea is to make ATM invisible to IP and the routers (routers 1 to 6 in this example). The ATM switches set up virtual circuits between themselves and the routers to create a fully meshed router network. The mesh is logical; there need not exist a physical link between each router. The ATM switches in the backbone are tasked with relaying the traffic. This idea is illustrated in Figure 6–8.

With *n* routers, there are $n \times (n-1)/2$ potential peer pairs in relation to routing and route advertising; this number can translate into substantial overhead. In Figure 6–8, six routers are placed on the ATM backbone, with the lines between the routers representing ATM virtual circuits. The potential peer relationships are $6 \times (5 - 1)/2 = 15$. The term

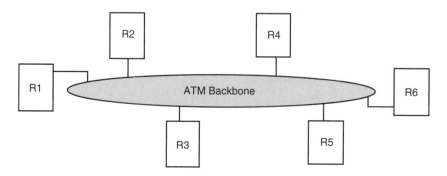

Figure 6–7   IP routers and ATM.

peer in this discussion refers to the idea of router neighbors and their adjacency to each other. In Internet routing, neighbor routers exchange information with each other about the address of which they are aware. If routing information is sent between the routers across each virtual circuit, then of course for a large network with many routers, the routing updates are going to consume a lot of the bandwidth of the network.

In Figure 6–9, four ATM switches (the boxes in the cloud) are linked to each other and to some of the routers. ATM switch A is directly connected to routers 1 and 2, switch B is connected to routers 4 and 6, switch C is connected to router 3, and switch D is connected to router 5. This approach is much more scalable in that the ATM switches are now running the IP routing protocols (thus, IP is overlaid onto the ATM switch). Routing adjacency is now a matter between the ATM switch and its directly attached router.

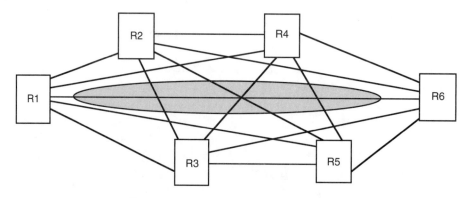

Figure 6–8   Fully meshed routers.

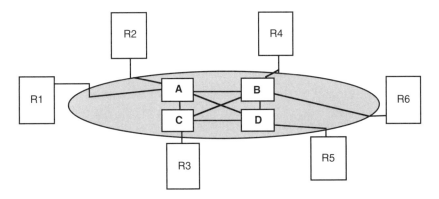

Figure 6–9   Running IP in ATM switches.

PRINCIPAL TRAFFIC MANAGEMENT ISSUES IN AND BETWEEN ATM-LSRs

It is obvious from the discussions so far in this chapter that ATM and MPLS switching have many similarities and differences. The major differences are:

- Syntaxes (and field sizes) of the VPI/VCI and MPLS label
- The inability of ATM to support VPI/VCI hierarchical stacking
- The small, fixed length ATM cell vs. the small-to-large, variable-length MPLS packet
- Upstream label allocation (ATM) vs. upstream or downstream label allocation (MPLS)
- The absence in MPLS (thus far) of QOS parameters such as the identification of class of traffic, variation tolerances, etc.

MPLS and ATM Interworking

These differences must be taken into account when MPLS and ATM are interworked. Furthermore, even though the number of bits allotted for the ATM VPI/VCI is quite large (24 bits), in practice many vendors have not built their switches to support more than 1000–4000 VPIs/VCIs per interface. This range of numbers is still a lot of virtual circuits per interface, but in large networks they are not considered an unlimited resource. In contrast, MPLS provides 20 bits for the label (per platform or per interface), which permits a much larger range of values for MPLS labels than exist for ATM VPIs/VCIs.

Difficulty of Interleaving AAL5 Traffic

After the release of the original ATM specification by the ITU-T, the AAL part of the standard came under fire. At that time, four AAL types had been defined, AAL1 through AAL4 (AAL2 was not fully defined until later[1]). AAL4 permitted the interleaving of multiple VPI/VCI cells across one link to the end user because it used sequence numbers and a multiplexing ID in the AAL header. These fields allowed the receiver to distinguish different users that used a common VPI/VCI.

However, AAL4 ATM vendors did not support AAL4 to any significant extent because of the large amount of overhead residing in the 48-byte service data unit (SDU). Consequently, revisions to ATM combined AAL3 and AAL4, renaming it AAL3/4. Revisions also added a new type called AAL5.

AAL5 has become the prevalent AAL option in the industry because it is quite efficient and consumes very little overhead. However, AAL5 has no sequence number field for any ID that can distinguish a 48-byte SDU from another SDU. Therefore, it is not possible to interleave different AAL5 SDUs with the same VPI/VCI to an end-user application; the application cannot know how to demultiplex (separate) the traffic to its constituent users. So, AAL5 traffic (all SDUs) belonging to a VPI/VCI must be sent contiguously before another SDU is sent. A bit in the ATM cell informs the receiver that the end of the SDU has been reached and it can expect the next cell to contain a different SDU.

The idea is illustrated in Figure 6–10. Cells with VPI/VCI values of 12 and 14 are sent to node C from node A and B respectively. The symbol of "v" above cells 12 and some of cells 40 are used in the figure to show that these cells originated at A. The other cells came from node B.

Node C's cell cross-connect table has been set up to map VPIs/VCIs 12 and 14 into VPI/VCI 40. That being the case, node D has no way of distinguishing node A's cells from those of node B, since AAL5 has no way to identify specific SDUs for different users.

---

[1]AAL2 has multiplexing/interleaving capabilities. It is being deployed in some systems that multiplex voice, video, and data of the same application onto the same VPI/VCI.

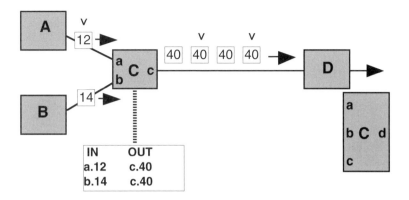

Figure 6–10   Losing traffic identity.

VC AND VP MERGING

MPLS handles the situation of SDU ambiguity with two options, VP merge and VC merge. Let's use VC merge to explain the idea, as implemented in Cisco LSRs (which is really the same as that described above for ATM VPIs/VCIs). The ATM-LSR is modified slightly to make certain that multiple cells coming together into one VC are not interleaved. These nodes simply buffer the incoming cells until they encounter the end of SDU bit in the cell header; then, they send the cells on to the next node. This process is called a VC merge, and is shown in Figure 6–11, along with some other operations that are explained next.

The incoming traffic consists of MPLS labels, and they are merged into ATM VCIs. The idea of VC merging is for node D to process all SDUs for 12, then all SDUs for 13, and then all SDUs for 14, never interleaving these SDUs.

The Merging Process

As Figure 6–11 shows, many MPLS labels can be merged into far fewer ATM VCIs (or VPIs for that matter), although this example shows only three labels in the merging operation. This approach conserves the limited reservoir of ATM "labels."

Let's examine Figure 6–11 in more detail. Starting at the left side of the figure:

- Nodes A, B, to X pass MPLS packets to node C, an ATM-LSR. The ATM-LSR MPLS/ATM cross-connect table correlates labels 12,

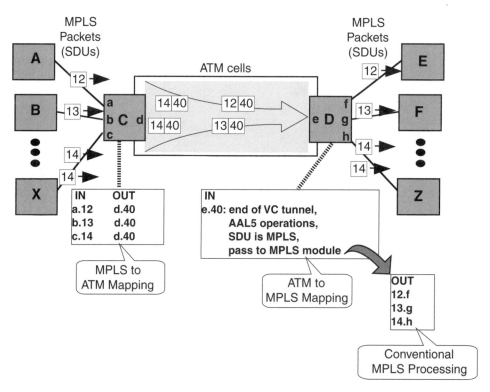

Figure 6–11   Example of VC merging, ATM backbone, and VC demerging.

13, and 14 on interfaces a, b, and c, respectively, with VC 40 on interface d.

- Node C performs AAL5 operations on the MPLS packets, perhaps dividing each MPLS packet into multiple 48-byte AAL SDUs. The MPLS label (a) can remain intact in the payload of the SDU, (b) can be stripped and placed into the ATM cell VCI field, (c) or both (a) and (b). These scenarios are discussed in "Mapping the MPLS Labels to ATM VPIs/VCIs." In any case, this operation completes the MPLS to ATM mapping functions at node C.

- The ATM cells are transported across the ATM backbone where the MPLS packets are not examined (depending on the specific ATM node). The ATM cell header's VPI/VCI makes relaying decisions about the next node in the network.

- The ATM cells arrive at ATM-LSR D. This node is configured to know that VPI/VCI 40 arriving on interface e is the end of the

ATM virtual circuit. Therefore, it strips away the 5-byte ATM cell header and passes the 48-byte SDU to AAL5.

- AAL5 at node D reassembles the SDUs into their full MPLS packets. The encapsulation field in the SDU header reveals that the payload is MPLS traffic. Therefore, node D's operating system passes these packets to MPLS. This operation completes node D's ATM-to-MPLS mapping functions.
- MPLS at node D examines the label in each packet and uses its MPLS LFIB (shown at the bottom right of Figure 6–11) to send the traffic to the next node across interfaces f, g, and h. This operation completes node D's MPLS functions.

Several observations are worth noting about the operations in Figure 6–11. First, only one ATM VPI/VCI has been expended to support three MPLS LSPs. Second, an ATM backbone is being employed to carry the MPLS packets from MPLS networks on the left side of the figure to MPLS networks on the right side of the figure. Thus, the current prevalent ATM backbone technology can be used with alterations made only to ingress and egress ATM nodes. The core ATM nodes do not need to execute MPLS (and AAL) operations. Third, the MPLS labels of 12, 13, and 14 need not be altered (swapped) in this operation, because the ATM backbone (and nodes C and D) is behaving like a point-to-point link between the MPLS user nodes. Consequently, node A/E, B/F, and X/Z operate as MPLS peer nodes and bind labels 12, 13, and 14 between them. Nodes C and D act as pass-through nodes for these three LSPs.

Of course, the ATM nodes must be configured to know how to encapsulate (at node C) and decapsulate (at node D) the MPLS packets. For these operations, the following thoughts are pertinent:

The ATM and MPLS operations at node D require the ATM cells coming in on interface e to be mapped to interfaces f, g, and h. This operation can be accomplished by the ATM/LSR cross-connect table and software being set up to strip off the ATM header, examine the MPLS label, and send the packet to the appropriate interface.

Reversing the Process with an MPLS Backbone

The process in Figure 6–11 can be reversed. MPLS can act as the backbone to outlying ATM networks. For this topology, the ATM cells are tunneled through the MPLS nodes. Node C becomes the MPLS ingress node and encapsulates the cells into the MPLS packets. Node D is the MPLS egress node and decapsulates the traffic back to native ATM cells.

The idea of label to VP or VC merging is quite simple, but like most aspects of interworking different protocols, the devil is in the details. The next part of this chapter explains the details. If you are not interested in a more detailed discussion of VP and VC merging, skip to the "Aspects of Frame Relay of Interest to MPLS" section of this chapter.

MAPPING THE MPLS LABELS TO ATM VPIs/VCIs

Figure 6–12 shows three methods for mapping MPLS labels into the ATM VPI/VCI fields. With method A, called SVC encoding, the top label in the stack is encoded into the VPI and VCI fields. Each LSP is correlated with an ATM SVC. ATM's Q.2931 can be used to act as the label distribution protocol. This method is simple, but there is no way to push or pop a label stack, since the ATM cell header has only one field (the combined VPI/VCI) to hold a label.

With method B, called SVP encoding, the top label in the stack is mapped to the VPI field and the second label is mapped into the VCI field. Obviously, two labels are supported, and in addition, the VPI can be used to support VP switching at the ATM-LSRs. For this method, the ATM-LSR at the egress of the VP does a pop operation.

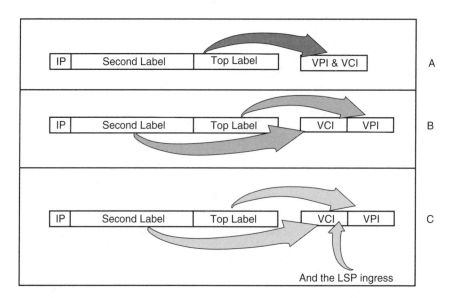

Figure 6–12   Three methods of mapping MLPLS labels to ATM VPI/VCIs.

With method C, called SVP multipoint encoding, the top label in the stack is mapped to the VPI field, and the second label is mapped into part of the VCI field. The remainder of the VCI field is used to identify the LSP ingress LSR.

With this technique, conventional ATM VP-switching capabilities can be used to provide multipoint-to-point VPs, thereby allowing cells from different packets to carry different VCI values.

TYPES OF MERGING (OR NONMERGING)

Several options are available in dealing with MPLS and ATM labels.

- VP merge
- Non-VC merge
- VC merge with no cell interleaving

VP merge entails the use of different VCIs within a VP to distinguish different sources. Non-VC merging simply uses a VC to identify each user cell stream.

The third approach is called VC merge with no cell interleave. It requires the LSR to buffer cells coming in from one packet until the complete packet has been received (as determined by an examination of the AAL5 end of the frame indicator). With VC merge with no cell interleaving, the LSR maps incoming VC labels for the same destination to the same outgoing VC label. When VC merge is used, switches are required to buffer cells, so this approach does incur delays at merge points. That is, cells belonging to different packets going to the same destination cannot be interleaved (recall that this restriction is because of AAL5).

INTEROPERATION OF VC MERGE, VP MERGE, AND NONMERGE

The interoperation of MPLS and ATM with regard to label merging or nonmerging is defined in section 26 of [ROSE01c].

First, in the situation where VC merge and nonmerge nodes are interconnected, the forwarding of cells is based in all cases on a VC (using both the VPI and VCI fields). If an upstream neighbor is doing a VC merge, then that upstream neighbor requires only a single VPI/VCI for a particular stream. If the upstream neighbor is not doing a merge, then the neighbor will require a single VPI/VCI per stream for itself, plus enough VPI/VCIs to pass to its upstream neighbors.

Second, in the situation with the VP merge node, rather than requesting a single VPI/VCI or a number of VPI/VCIs from its downstream neighbor, the LSR may request a single VP (identified by a VPI), but several VCIs within the VP. The packets/cells associated with the VPI all go to the same destination. The use of VCs within the VP permits cell interleaving.

Suppose that a nonmerge node is downstream from two different VP merge nodes. This LSR may need to request one VPI/VCI (for traffic originating from itself) plus two VPs (one for each upstream node), each associated with a specified set of VCIs (as requested from the upstream node).

To support all of VP merge, VC merge, and nonmerge, upstream nodes must be allowed to request a combination of zero or more VC identifiers (consisting of a VPI/VCI), plus zero or more VPs (identified by VPIs), each containing a specified number of VCs (identified by a set of VCIs, which are significant within a VP). VP merge nodes would therefore request one VP, with a contained VCI for traffic that it originates (if appropriate) plus a VCI for each VC requested from above (regardless of whether or not the VC is part of a containing VP). VC merge node would request only a single VPI/VCI (since they can merge all upstream traffic into a single VC). Nonmerge nodes would pass on any requests that they get from above and would request a VPI/VCI for traffic that they originate (if appropriate).

Non-VC merging is simple and allows the receiving node to easily reassemble cells into packets, since the VC values distinguish the senders. However, each LSR must keep track of each VC label for n sources and destinations for a fully meshed connectivity. This situation presents a scaling problem, since the LSR must manage $o(n^2)$ labels; so, for 1,000 sources/destinations, the VC routing table is 1,000,000 entries.

For VP merging, each LSR now manages $o(n)$ VP labels, clearly an attractive alternative to non-VC merging. But this approach requires more processing on the part of the node to keep track of VPs and their associated VCs.

THE VIRTUAL CIRCUIT ID

The VPIs and VCIs in the ATM cell change as the cell is relayed from node to node. Unlike the case with MPLS and LDP, there is no end-to-end identifier (IP address) in the cell. Therefore, the ATM cell cannot be used to identify a VC end-to-end. For MPLS to be used on ATM links, the ATM VCs must be identified in the LDP mapping messages. For this purpose, the virtual connection ID (VCID) is used. It has the same value at both ends of the VC.

Two categories of VCID notification procedures are defined.

- The inband procedure: the notification messages are forwarded over the VC to which they refer.
- The out-of-band procedure: the notification messages are forwarded over a different VC to which they refer.

NOTIFICATION OPERATION

Figure 6–13 shows an example of an inband notification operation. The node A establishes a VC to the destination node B (by signaling or management). Next, node A selects a VCID value. Then, node A sends a VCID PROPOSE message, which contains the VCID value and a message ID, through the newly established VC to node B.

Node A establishes an association between the outgoing label (VPI/VCI) for the VC and the VCID value. Node B receives the message from the VC and establishes an association between the VCID in the message and the incoming label (VPI/VCI) for the VC. Until node B receives the LDP REQUEST message, node B discards any packet received over the VC other than the VCID PROPOSE message.

Node B sends an acknowledgment (ACK) message to node A. This message contains the same VCID and message ID as specified in the received message.

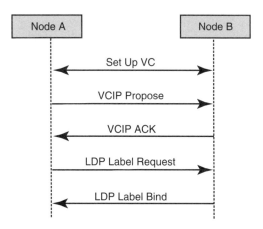

Figure 6–13   Inband notification operation.

When node A receives the ACK message, it checks whether the VCID and the message ID in the message are the same as the registered ones. If they are the same, node A regards that node B has established the association between the VC and VCID. Otherwise, the message is ignored. If node A does not receive the ACK message with the expected message ID and VCID during a given period, node A resends the VCID PROPOSE message to node B.

Finally, after receiving the ACK message, node A sends an LDP REQUEST message to node B. It contains the message ID used for VCID PROPOSE. When node B receives the LDP REQUEST message, it assumes node A has received the ACK correctly.

The message exchange with VCID PROPOSE, VCID ACK, and LDP REQUEST messages constitutes a three-way handshake. The three-way handshake mechanism is required since the transmission of VCID PROPOSE message is unreliable. Once the three-way handshake is completed, node B ignores all VCID PROPOSE messages received over the VC.

The out-of-band operation is similar, except (typically) an ATM signaling message (Q.2931) carries the VCID. Consequently, the PROPOSE and PROPOSE ACK messages are replaced by the ATM signaling messages.

VPI/VCI VALUES

When two LSRs are directly connected through an ATM interface, they jointly control the allocation of VPIs/VCIs on this interface. They may agree to use the VPI/VCI field to encode a single label. The default VPI/VCI value for the non-MPLS connection is VPI 0, VCI 32. Other values can be configured, as long as both parties are aware of the configured value.

A VPI/VCI value whose VCI value is in the range 0 to 32 cannot be used as the encoding of a label. With the exception of these reserved values, the VPI/VCI values used in the two directions of the link can be treated as independent spaces. The allowable ranges of VCIs are communicated through LDP.

ENCAPSULATION AND TTL OPERATIONS

Labeled packets must be transmitted by means of the null encapsulation of Section 5.1 of RFC 1483. Except in certain circumstances specified below, when a labeled packet is transmitted on an LC-ATM interface, where the VPI/VCI (or VCID) is interpreted as the top label in the label stack, the packet must also contain a shim header.

If the packet has a label stack with n entries, it must carry a shim header with n entries. The actual value of the top label is encoded in the VPI/VCI field. The label value of the top entry in the shim (which is just a "placeholder" entry) *must* be set to 0 upon transmission and must be ignored upon reception. The packet's outgoing TTL and its class of service (CoS) are carried in the TTL and CoS fields, respectively, of the top stack entry in the shim.

We now turn our attention to MPLS and Frame Relay. We will see that many of the operations of MPLS and ATM are similar to those of MPLS and Frame Relay.

ASPECTS OF FRAME RELAY OF INTEREST TO MPLS

This section provides an overview of those aspects of Frame Relay that are relevant to MPLS. [CONT98] represents the work from the IETF on Frame Relay and MPLS. These topics are of interest:

- Virtual circuits: the logical connections in the network
- DLCIs: the Frame Relay labels (virtual circuit IDs)
- Frame Relay header: contains the labels
- Permanent virtual circuits (PVCs) and switched virtual calls (SVCs)

Virtual Circuits and DLCIs

Frame Relay virtual circuits are similar to ATM virtual circuits, with one major difference. Frame Relay uses only one value, called the data link connection identifier (DLCI), to identify a VC. These ideas are illustrated in Figure 6–14. Routing in the Frame Relay is quite similar to that

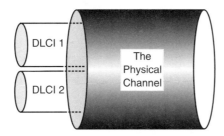

Figure 6–14 Frame relay virtual circuits and DLCIs.

in an ATM network. It is performed by the Frame Relay switch examining the DLCI field in the Frame Relay frame to make a forwarding decision.

The Frame Relay Header

The Frame Relay frame resembles many other protocols that use the High-Level Data Link Control (HDLC) frame format, illustrated in Figure 6–15. It contains the beginning flag used to delimit and recognize the frame on the communications link. The ending flag signals the beginning of the next frame. Frame Relay does not contain a separate address field; the address field is contained in the control field. Together they are designated as the Frame Relay header. The information field contains user data, such as TCP/IP traffic. The frame check sequence (FCS) field, as in other link layer protocols, is used to determine if the frame has been damaged during transmission over the communications link.

The Frame Relay header consists of six fields. They are listed and briefly described here and explained in more detail in subsequent discussions.

- *DLCI.* The data link connection identifier identifies the virtual circuit user (which is typically a router attached to a Frame Relay network but can be any machine with a Frame Relay interface).
- *C/R.* The command response bit (not used by Frame Relay).
- *EA.* The address extension bits, shown in bit positions 1 in the two octets of the header. Used to extend the length of the DLCI.
- *FECN.* The forward explicit congestion notification bit.
- *BECN.* The backward explicit congestion notification bit.
- *DE.* The discard eligibility indicator bit.

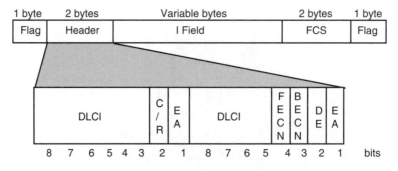

Figure 6–15   Frame relay PDU (frame).

Permanent Virtual Circuits and Switched Virtual Calls

A virtual circuit can be provisioned as a permanent virtual circuit (PVC) or a switched virtual call (SVC) and is identical to the ATM operations explained earlier. The difference is that the DLCI identifies the virtual circuit.

Some of the key features of Frame Relay switches that affect their behavior as LSRs are the following:

• The label swapping function is performed with the DLCI. The function is similar to the ATM operation, except Frame Relay has only one value for the label.

• There is no capability to perform a TTL-decrement function as is performed on IP headers in routers.

• Congestion control is performed by each node according to parameters that are passed at circuit creation. The BECN and FECN bits in the frame headers may be set as a consequence of congestion or of the contractual parameters of the circuit being exceeded.

RUNNING ATM, FRAME RELAY AND OTHERS OVER MPLS

In anticipation of the future, where native-mode MPLS networks are in operation, the IETF has been working on specifications for running layer 2 protcols such as ATM and Frame Relay over MPLS. Most of the issues deal with encapsulation headers and segmentation/reassembly procedures. I refer you to [MART01] if you wish to pursue this aspect of MPLS.

SUMMARY

For the foreseeable future, ATM and Frame Relay will be prevalent protocols in wide area networks. This chapter explained the major features of ATM and Frame Relay and summarized the ways in which ATM and Frame Relay bearer services can support a limited set of MPLS operations. The principal problems are the inability of ATM and Frame Relay to support deep label stacks and the absence of bits to handle TTL operations.

7

Traffic Engineering

This chapter discusses three aspects of MPLS. The first aspect is how networks in general are engineered to provide efficient services to their customers. The second aspect is how MPLS plays a role in supporting these services. The third aspect deals with protection switching: how LSRs and LSPs can be engineered to provide alternate routes through the MPLS routing domain if nodes and or links fail.

The chapter explains traffic classes and traffic engineering tools to manage these classes, including policing with token and leaky buckets, traffic shaping with different kinds of queue service algorithms, and induced MPLS graphs. We conclude with examples of how weighted fair queuing (WFQ) is used to manage MPLS flows. To see how MPLS traffic engineering can be correlated with ATM, Frame Relay, and RSVP, see Appendix B.

TRAFFIC ENGINEERING DEFINED

Traffic engineering (TE) deals with the performance of a network in supporting the network's customers and their QOS needs. The focus of TE for MPLS networks is (a) the measurement of traffic and (b) the control of traffic. The latter operation deals with operations to ensure that the network has the resources to support the users' QOS requirements.

The Internet Working Group [AWDU99] has published RFC 2702. This informational RFC defines in a general way the requirements for traffic engineering over MPLS. The next part of this chapter provides a summary of [AWDU99], with the author's tutorial comments added to the discussion.

TRAFFIC-ORIENTED OR RESOURCE-ORIENTED PERFORMANCE

Traffic engineering in an MPLS environment establishes objectives with regard to two performance functions: (a) traffic-oriented objectives and (b) resource-oriented objectives.

Traffic-oriented performance supports the QOS operations of user traffic. In a single-class, best-effort Internet service model, the key traffic-oriented performance objectives include minimizing traffic loss, minimizing delay, maximizing throughput, and enforcing service level agreements (SLAs).

Resource-oriented performance objectives deal with the network resources, such as communications links, routers, and servers—those entities that contribute to the realization of traffic-oriented objectives.

Efficient management of these resources is vital to the attainment of resource-oriented performance objectives. Available bandwidth is the bottom line; without bandwidth, any number of TE operations are worthless, and the efficient management of the available bandwidth is the essence of TE.

THE CONGESTION PROBLEM

Any network that admits traffic and users on demand (such as an internet) must deal with the problem of congestion. The management of all users' traffic to prevent congestion is an important aspect of the QOS picture. Congestion translates into reduced throughput and increased delay. Congestion is the death knell of effective QOS.

Problem Description

Most networks provide transmission rules for their users, including agreements on how much traffic can be sent to the network before the traffic flow is regulated (flow controlled). Flow control is an essential ingredient to prevent congestion in a network. It is easy to understand the concern network managers have about congestion, because it can result in severe degradation of the network operations, both in throughput and response time.

As the traffic (offered load) in the network reaches a certain point, mild congestion begins to occur, with the resulting drop in throughput. Figure 7–1 depicts the problem. If this situation proceeded in a linear fashion, it would not be so complex a problem. However, at a time when utilization of the network reaches a certain level, throughput drops precipitously because of serious congestion and the buildup of packets at the servers' queues.

Therefore, networks must (a) provide some mechanism of informing affected nodes inside the network when congestion is occurring and (b) provide a flow control mechanism on external user devices outside the network.

Two Scenarios of Congestion

Minimizing congestion is one of the most important traffic and resource-oriented performance objectives. In this discussion, we address the situation shown in Figure 7–1, wherein the congestion lasts for a prolonged period, and not one in which a short-lived burst of traffic momentarily exceeds the bandwidth capability of the network.

With this assumption, congestion can be described in two scenarios. The first is straightforward: there are insufficient resources to accommodate the user's traffic. The second is considerably more complex: there are sufficient resources in the network to support the users' QOS needs, but the traffic streams are not mapped properly onto the available

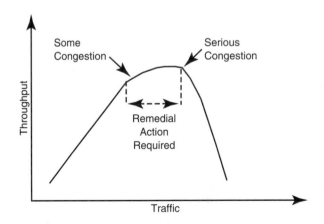

Figure 7–1   Potential congestion problems.

network resources (principally, the communications links betw. nodes). Therefore, some parts of the network become underutilized and others are saturated with user traffic.

We can solve the first problem by building networks with more bandwidth (say, in a freeway analogy, putting in more freeways). We also can help matters by applying congestion control techniques, such as window control operations with "receive not ready," and "congestion notification" (in the freeway analogy, placing traffic lights at the entrance ramps to the freeway). The major problem with the "more bandwidth" philosophy is that it leads to very poor utilization of very expensive network resources during periods when there is less traffic (say, during the early morning hours). It is akin to building a freeway system that accepts all rush hour traffic, and at 2:00 a.m., the 20 lanes of asphalt are almost empty).[1]

The second type of problem, inefficient resource allocation, can usually be addressed through traffic engineering. After all, the resources are available in the network. It is a matter of finding them and diverting user traffic to them.

In general, congestion resulting from inefficient resource allocation can be reduced by the adoption of load-balancing policies; that is, diverting traffic to available links and nodes. The idea is to minimize maximum congestion by avoiding that unfortunate curve shown in Figure 7–1. Obviously, the result is increased throughput and decreased lost and delayed traffic.

SERVICES BASED ON QOS NEEDS AND CLASSES OF TRAFFIC

Traffic can be organized around a concept called service classes, which are summarized in Table 7–1. These traffic classes are similar to those used in ATM networks. The classes are defined with respect to the following operations:

- Timing between sender and receiver (present or not present)
- Bit rate (variable or constant)
- Connectionless or connection-oriented sessions between sender and receiver

---

[1]An important footnote to this discussion is the view by some people that the building of networks with adequate bandwidth to support any set of users is possible. With the advent of WDM and optical switches, it is felt that the concern with bandwidth management will become a moot point. It is a provocative idea; I await the results.

Table 7–1   Traffic Engineering Service Classes

Class	Features
Class A	Constant bit rate (CBR), TDM-based
	Connection-oriented
	Timing required, flow control must be minimal
	Some loss permitted
Class B	Variable bit rate (VBR), STDM-based (bursty)
	Connection-oriented
	Timing required, flow control must be minimal
	Some loss permitted
Class C	Variable bit rate, STDM-based (bursty)
	Connection-oriented
	Timing not required, flow control permitted
	No loss permitted
Class D	Variable bit rate, STDM-based (bursty)
	Connectionless
	Timing not required, flow control permitted
	No loss permitted

- Sequencing of user payload
- Flow control operations
- Accounting for user traffic
- Segmentation and reassembly (SAR) of user PDUs (protocol data units)

The acronyms TDM and STDM in Table 7–1 refer to time division multiplexing and statistical time division multiplexing, respectively. TDM provides a predictable level of service; the user is provided with a constant bit rate flow and periodic slots on the channels, typically every 125 microseconds. STDM provides no periodic slots. The user is given bursts of time on the channel.

TRAFFIC ENGINEERING AND TRAFFIC SHAPING

In its simplest form, traffic engineering attempts to optimize users' QOS needs by making the best use of network resources to support those needs. The limitation, of course, is the network resources. Therefore, with

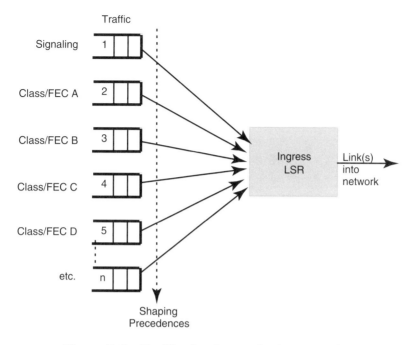

Figure 7–2   Traffic shaping at the ingress LSR.

a network that does not have sufficient bandwidth (link speeds and LSR processing power) to support all users' QOS requirements every moment of the day, traffic engineering operations must "shape" the users' traffic. This means that mechanisms must be in place to determine (shape) how the network supports the different classes of user traffic. As depicted in Figure 7–2, the shaping should occur at the ingress LSR and entails setting up queues and acting on priorities assigned to the traffic class (giving precedence to one class or FEC over another). In this example, each traffic class is assigned to a different queue, and signaling traffic is given precedence over the other traffic classes, which is a common practice.

QUEUING THE TRAFFIC

Many systems in place today, especially routers, support several types of queues. The prevalent types are listed next.

- First-in, first-out queuing (FIFO): Transmission of packets is based on their order of arrival. MPLS uses this method for a given FEC.

- Weighted fair queuing (WFQ): The available bandwidth across queues of traffic is divided according to weights. Given its weight, each traffic class is treated fairly. This approach is often used when the overall traffic is a mix of multiple traffic classes. Class A traffic is accorded a heavier weight than, say, class D traffic. WFQ is well suited to manage MPLS flows, as described in more detail in the "Examples of WFQ and MPLS Flows" section of this chapter.
- Custom queuing (CQ): Bandwidth is allotted proportionally for each traffic class. CQ guarantees some level of service to all traffic classes.
- Priority Queuing (PQ): All packets belonging to a higher-priority class are transmitted before any lower-priority class. Therefore, some traffic is transmitted at the expense of other traffic.

PROBLEMS WITH EXISTING ROUTING OPERATIONS

In current internets, the routes between sending and receiving parties are set up by routing protocols, such as OSPF and BGP. These protocols are not designed for route and resource optimization, since they are based on shortest path (the fewest number of hops) operations. They build a route according to the topology of the network, not according to the bandwidth of the network. In addition, they generally do not consider the class of traffic in establishing these routes. Thus, many networks must adapt schemes to load-balance links and nodes that have not been chosen by the routing protocols.

For example, in Figure 7–3, router A is informed by a routing protocol that it can "reach" host G through two possible paths, one that goes to C first or one that goes to B first. With a typical routing protocol, the path taken is through C because the path has the fewest number of hops to the destination. That path may indeed be the better one. But if the link between routers A and C becomes congested or if the link is not a high-capacity link (say, it is an OC-12 link in contrast to the links on the other path that are OC-48 links), then traffic will be concentrated on the A – B path. The A – C path goes underutilized.

Certainly, we can take measures to "force" traffic to parts of the network that are underutilized. But the communications links of most networks vary in their capacity, and it is a huge task to use conventional IP routing techniques to overcome the problem of under- or overutilization of parts of a network. In the complex Internet, IP and the Internet routing protocols have not proved adequate to this task.

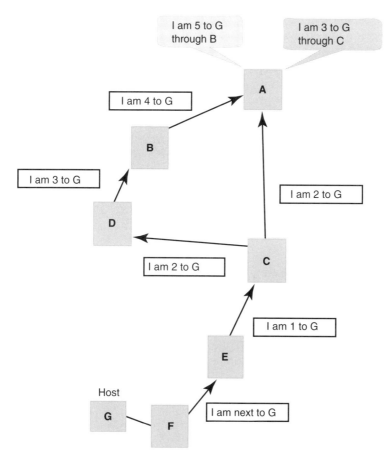

Figure 7–3   Current approach.

THE OVERLAY NETWORK APPROACH

A common approach to ameliorate the problem is to use an overlay network. This approach means running IP over an ATM or a Frame Relay bearer network and using ATM or Frame Relay virtual circuits as a mechanism to improve network resource utilization. For this discussion, the bearer network is ATM.

The routing protocols are still in operation and are still performing route advertising, but they are not being used (by themselves) to determine the route. By configuration of permanent virtual circuits (PVCs) across the paths, the overall operations of the network are improved, as are the services to the network users. In addition, ATM has several features (admission control into the network, policing of traffic, congestion notification, traffic shaping, virtual circuit rerouting, as examples) that provide powerful QOS operations.

Then why not use ATM? The answer is that some engineers think ATM is ill-designed from the standpoint of its consumption of bandwidth (a small payload size of 48 bytes versus a relatively large header of 5 bytes). Others do not like the connection-oriented nature of ATM, whereby a PVC must be set up before any traffic can be sent. Alternately, a connection-on-demand must be made for traffic coming in on-the-fly, with an ATM switched virtual call (SVC).

These points are well made, but the fact remains that as the Internet evolves and as the many Internet QOS and label switching RFCs are approved, the Internet will take on many of the characteristics of an ATM network.

One of the goals of an MPLS-based network is to exhibit the power of an ATM network but not be constrained to the small cells and other vexing attributes, such as ATM's optimal error-checking design for optical fiber links, its inherent use of a label hierarchy (two-level), and its graceful OAM operations with SONET.

Obviously, I am being a bit facetious with my last comments. ATM is an excellent technology, but it is expensive to implement. Alternatives, such as MPLS, look to perform like an ATM network at a lesser cost.

INDUCED MPLS GRAPH

An induced MPLS graph is analogous to a virtual circuit *logical* topology in the overlay network just discussed. The induced MPLS graph is a set of LSRs (the nodes of the graph) and a set of LSPs that together provide the *logical* connectivity between the LSRs. As discussed in Chapter 4, it is possible to use label stacks to create "hierarchical-induced MPLS graphs."

[AWDU99] establishes the rationale for induced MPLS graphs; the abstract description of these graphs is as follows.

> Induced MPLS graphs are important because the basic problem of bandwidth management in an MPLS domain is the issue of how to efficiently map an induced MPLS graph onto the physical network topology. The induced MPLS graph abstraction is formalized below.
>
> Let G = (V, E, c) be a capacitated graph depicting the physical topology of the network. Here, V is the set of nodes in the network and E is the set of links; that is, for v and w in V, the object (v,w) is in E if v and w are directly

connected under G. The parameter "c" is a set of capacity and other constraints associated with E and V. We will refer to G as the "base" network topology.

Let H = (U, F, d) be the induced MPLS graph, where U is a subset of V representing the set of LSRs in the network, or more precisely, the set of LSRs that are the endpoints of at least one LSP. Here, F is the set of LSPs, so that for x and y in U, the object (x, y) is in F if there is an LSP with x and y as endpoints. The parameter "d" is the set of demands and restrictions associated with F. Evidently, H is a directed graph. It can be seen that H depends on the transitivity characteristics of G.

TRAFFIC TRUNKS, TRAFFIC FLOWS, AND LABEL SWITCHED PATHS

An important aspect of MPLS TE is the distinction of traffic trunks, traffic flows, and LSPs. A traffic trunk is an aggregation of traffic flows of the same class that are placed inside an LSP. A traffic trunk can have characteristics associated with it (addresses, port numbers). A traffic trunk can be routed because it is an aspect of the LSP. Therefore, the path through which the traffic trunk flows can be changed.

MPLS TE concerns itself with mapping traffic trunks onto the physical links of a network through label switched paths. Stated another way, an induced MPLS graph (R) is mapped onto the physical network topology (G).

ATTRACTIVENESS OF MPLS FOR TRAFFIC ENGINEERING

Now that some basic concepts have been explained, we can see that an MPLS-based network lends itself to TE operations because of the following characteristics:

- Label switches are not constrained to the conventional IP forwarding dictated by conventional IP-based routing protocols.
- Traffic trunks can be mapped onto label switched paths.
- Attributes can be associated with traffic trunks.
- IP forwarding permits only address aggregation, whereas MPLS permits aggregation or disaggregation.
- Constraint-based routing is relatively easy to implement.
- MPLS can be implemented at less cost than ATM.

LINK CAPACITY: THE ULTIMATE ARBITER

The MPLS TE model takes into account the fact that the bandwidth capacity of the links in a network is the ultimate arbiter of traffic engineering decisions.[2] To see why this is so, consider that a SONET OC-3 link with a 155.52 Mbit/s capacity can accept an absolute maximum of 353,207 ATM cells per second. Given the assumption that the system is consuming the OC-3 bandwidth perfectly, the following holds: 155,520,000 (less the overhead of the 155.52 Mbit/s OC-3 frame yields a rate of 149.760 Mbit/s)/ 424 bits in a 53-octet cell = 353,207 cells per second.

No more traffic than the 353k cells per second can be placed on this physical link. Consequently, it is the task of traffic engineering to make efficient use of this link with one of the following operations: (a) move traffic away from this link interface if the traffic rate exceeds the link bandwidth rate, or (b) place more traffic onto this link interface if the traffic rate is less than the link bandwidth rate.

LOAD DISTRIBUTION

In many situations, it makes sense to distribute the traffic across parallel traffic trunks or, in some situations, across diverse paths and LSRs in the physical network. For the first operation (see Figure 7–4), multiple traffic trunks are set up between two adjacent nodes (the links between

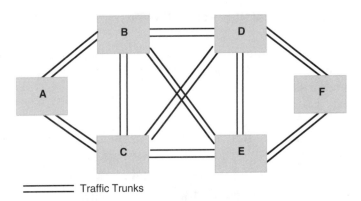

Figure 7–4   Load distribution.

---

[2]This statement is made with the assumption that the nodes (routers) in the network have the capacity to service the links attached to the routers.

LSRs A – F), allowing each traffic trunk to carry a portion of the total traffic load. This approach is quite common and today is implemented in many networks, such as SS7.

The second operation, distributing the traffic load on diverse trunks and different nodes in the network, is a more complex problem and is addressed in the "Examples of MPLS Protection Switching" section of this chapter.

TRAFFIC TRUNK ATTRIBUTES

MPLS TE establishes the following attributes for traffic trunks.

1. In accordance with the idea of the FEC, a traffic trunk is an aggregate of traffic flows belonging to the same class, although this attribute is not cast in stone. For example, it may be desirable to place different classes of traffic into an FEC if detailed traffic granularity is not needed.
2. A traffic trunk can encapsulate an FEC between any ingress LSR and an egress LSR.
3. Traffic trunks, through the FEC label, are routable.
4. A traffic trunk can be moved from one path to another, which means it is distinct from the LSP through which it travels.
5. A traffic trunk is unidirectional, but in practice, two of these trunks can be associated with each other as long as they are created and destroyed together. This association is called a bidirectional traffic trunk (BTT). The two BTTs do not have to traverse the same physical paths, although it may be desirable that they do if the two flows are tightly coupled with regard to real-time interaction. For example, if one traffic trunk traverses through many more LSRs than its partner, it might affect the quality of the interaction. In any event, a BTT is called *topologically symmetric* if the traffic trunks are on the same physical path and *topologically asymmetric* if they are routed through different physical paths.

Attributes of Traffic Trunks for Traffic Engineering

A traffic trunk has parameters assigned to it that identify its attributes. In turn, these attributes influence its behavioral characteristics, that is, how the traffic is treated by the network. The attribute values are assigned by network administration (provisioned) or by software that automatically examines the FEC criteria (addresses, port numbers, and PID) and sets up the parameters.

The important traffic trunk attributes for traffic engineering are listed here and described in more detail in this section of the chapter. Some of these attributes are similar to ATM traffic engineering operations, and we show these similarities as well.

- Traffic parameter attribute
- Policing attribute
- Generic path selection and maintenance attributes
- Priority attribute
- Preemption attribute
- Resilience attribute

Traffic Parameter and Policing Attributes. These two attributes are grouped together because of their close relationships. They are similar to ATM's usage parameter control (UPC). Both MPLS TE and ATM UPC capture the FEC of the traffic, monitor and control traffic, and check on the validity of the traffic entering the network at the ingress node. ATM UPC maintains the integrity of the network and makes sure that only valid VPIs and VCIs are "entering" the network. For MPLS, this equivalent operation would entail monitoring FECs and associated labels.

Several other features are desirable for these attributes:

- The ability to detect noncompliant traffic
- The ability to vary the parameters that are checked
- A rapid response to the users that are violating their contract
- The ability to keep the operations of noncompliant users transparent to compliant users

The Generic Cell/Packet Rate Algorithm **(GC/PRA).** Many implementations (ATM and the Internet) use the ATM generic cell rate algorithm for implementing policing operations. Figure 7–5 shows the two algorithms available for the GC/PRA, which is implemented as a virtual scheduling algorithm or as a continuous-state leaky bucket algorithm. The two algorithms serve the same purpose: to make certain that packets or cells are either conforming (arriving within the bound of an

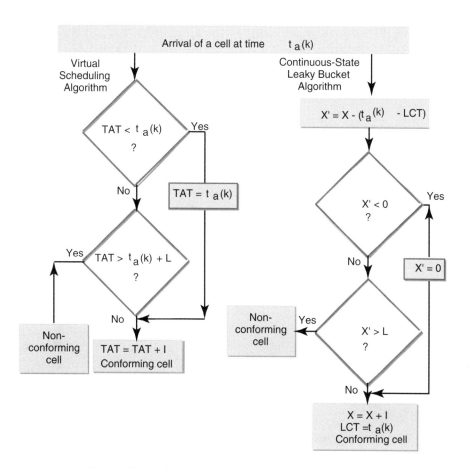

Figure 7–5   Generic cell rate algorithm (GCRA).

expected arrival time) or nonconforming (arriving sooner than an expected arrival time). The algorithm is applicable to cells, frames, packets, and any discrete protocol data unit. For ease of reading, I use the term *cell* hereafter.

First, two definitions are needed: the theoretical arrival time (TAT) is the nominal arrival time of the cell from the source, assuming that the source sends evenly spaced cells. Additionally, the parameter k is the kth cell in a stream of cells on the same virtual connection.

Given these definitions, the virtual scheduling algorithm operates as follows:

- After the arrival of the first cell, $t_a(1)$, the TAT is set to the current time; thereafter,

- If the arrival time of the kth cell is after the current value of TAT—in the flow chart: TAT < $t_a(k)$—then the cell is conforming and TAT is updated to $t_a(k)$ plus the increment I.
- If the kth cell's arrival time is greater than or equal to TAT – L but less than TAT—in the flow chart: TAT > $t_a(k)$ + L—then the cell is once again conforming and TAT is incremented to I.
- The cell is nonconforming if the arrival time of the kth cell is less than TAT–L—if TAT is greater than $t_a(k)$ + L. In this situation, the TAT is unchanged.

The continuous-state leaky bucket algorithm is viewed as a finite-capacity bucket whose content drains out at a continuous rate of 1 unit of content per time unit. Its content is increased by the increment I for each conforming cell. Simply stated, if at cell arrival, the content of the bucket is less than or equal to the limit L, the cell is conforming. Otherwise, it is nonconforming. The bucket capacity is L + I (the upper bound on the counter).

The continuous-state leaky bucket algorithm operates as follows:

- At the arrival of the first cell $t_a(1)$, the content of the bucket X is set to 0, and the last conformance time (LCT) is set to $t_a(1)$.
- At the arrival of the kth cell, $t_a(k)$, the content of the bucket is updated to the value X'. With this update, X' equals the content of the bucket X after the arrival of the last conforming cell minus the amount the bucket has drained since that arrival—in the flow chart: X' = X – $t_a(k)$ – LCT.
- The content of the bucket is not allowed to be negative—in the flow chart: X' < 0, then X' = 0.
- If X' is less than or equal to the limit value L, then the cell is conforming and the content of the bucket X is set to X' + I for the current cell and the LCT is set to current time $t_a(k)$.
- If X' is greater than L, then the cell is nonconforming and the values of X and LCT are not changed.

Generic Path Selection and Management Attribute.   This attribute is concerned with the selection of the route taken by the traffic trunk and the rules for the maintenance of paths that have been established. The paths can be derived from conventional routing protocols, such as OSPF or BGP, or they can be preconfigured (administratively specified explicit paths). For MPLS networks, a "path preference rule" is

associated with an administratively specified path and is set up as
mandatory or nonmandatory.

Resource class affinity attributes associated with a traffic trunk can
be used to specify the class of resources that are to be explicitly included
or excluded from the path of the traffic trunk. These policy attributes can
be used to impose additional constraints on the path traversed by a given
traffic trunk. Resource-class affinity attributes for traffic can be specified
as a sequence of tuples:

```
<resource-class, affinity>; <resource-class, affinity>; ..
```

The resource-class parameter identifies a resource class for which
an affinity relationship is defined with respect to the traffic trunk. The
affinity parameter indicates the affinity relationship; that is, whether
members of the resource class are to be included or excluded from the
path of the traffic trunk. The affinity parameter may be a binary variable
that takes one of the following values: (1) explicit inclusion or (2) explicit
exclusion.

An adaptivity attribute is a part of the path maintenance parameters associated with traffic trunks. The adaptivity attribute associated
with a traffic trunk indicates whether the trunk is subject to reoptimization. An adaptivity attribute is a binary variable that takes one of the following values: (1) permit reoptimization or (2) disable reoptimization.

Priority Attribute.   This attribute defines the relative importance
of traffic trunks and is quite important if constraint-based routing is
used in the network. This attribute is discussed in more detail later in
the "Examples of WFQ and MPLS Flows" section of this chapter and in
Chapter 8.

Preemption Attribute.   The preemption attribute determines
whether a traffic trunk can preempt another traffic trunk from a given
path and whether another trunk can preempt a specific traffic trunk.
Preemption is useful for both traffic-oriented and resource-oriented performance objectives. Preemption assures that high-priority traffic trunks
can always be routed through relatively favorable paths within a differentiated services environment. Preemption can also be used to implement various prioritized restoration policies following fault events.

These MPLS concepts are very similar to SS7 rules on its signaling
link selection rules. With MPLS, the preemption attribute can be used to
specify four preempt modes for a traffic trunk: (1) preemptor enabled,

(2) non-preemptor, (3) preemptable, and (4) non-preemptable. A preemptor-enabled traffic trunk can preempt lower-priority traffic trunks designated as preemptable. A traffic specified as non-preemptable cannot be preempted by any other trunks, regardless of relative priorities. A traffic trunk designated as preemptable can be preempted by higher-priority trunks that are preemptor enabled.

Some of the preempt modes are mutually exclusive. Using the numbering scheme depicted above, the feasible preempt mode combinations for a given traffic trunk are as follows: (1, 3), (1, 4), (2, 3), and (2, 4). The (2, 4) combination should be the default.

A traffic trunk, say A, can preempt another traffic trunk, say B, only if all of the following five conditions hold:

1. A has a relatively higher priority than B.
2. A contends for a resource used by B.
3. The resource cannot concurrently accommodate A and B based on certain decision criteria.
4. A is preemptor enabled.
5. B is preemptable.

Resilience Attribute   The resilience attribute determines the behavior of a traffic trunk under fault conditions; that is, when a fault occurs along the path traversed by the traffic trunk. The basic problems that need to be addressed under such circumstances are (1) fault detection, (2) failure notification, and (3) recovery and service restoration.

CONSTRAINT-BASED ROUTING (CR)

This section of the chapter explains the TE aspects of the constraint-based LDP (CR-LDP), explained in Chapter 9. For continuity of the two chapters, I have included some of this material in both chapters. Constraint-based routing (also called QOS routing and constrained routing) provides a route through the MPLS network, based on a user's QOS needs. It is demand driven and is aware of the traffic trunk attributes and the attributes of network resources. Each LSR automatically computes an explicit route for each traffic trunk, based on the requirements of the trunk's attributes and subject to the constraints of network resources and the administrative policies of the network.

ATM and Frame Relay networks have been using constraint-based routing for a number of years. Work is underway to extend these concepts for layer 3 operations. The focus is on extending OSPF and IS-IS to support constraint-based routing. For more information on using OSPF, see RFC 2676. Chapter 9 explains constraint-based routing operations using LDP. This part of the chapter explains the parameters that are exchanged between constraint-based LSRs (CR-LSRs) and how they are used by these LSRs.

Peak Rate

The peak rate defines the maximum rate at which traffic should be sent to the CR-LSP. The peak rate is useful for resource allocation. If resource allocation within the MPLS domain depends on the peak rate value, then it should be enforced at the ingress to the MPLS domain.

Committed Rate

The committed rate defines the rate at which the MPLS domain commits to be available to the CR-LSP.

Excess Burst Size

The excess burst size (EBS) can be used at the edge of an MPLS domain to condition the traffic. The excess burst size can be used to measure the extent by which the traffic sent on a CR-LSP exceeds the committed rate. The possible traffic conditioning actions, such as passing, marking, or dropping, are specific to the MPLS domain and are explained in Chapter 8.

Peak Rate Token Bucket

We examined the token bucket concept earlier. For MPLS and CR-LDP, it is defined as follows. The peak rate of a CR-LSP is specified in terms of a token bucket P with token rate PDR and maximum token bucket size PBS.

The token bucket P is initially (at time 0) full, i.e., the token count $Tp(0) = PBS$. Thereafter, the token count Tp, if less than PBS, is incremented by one PDR times per second. When a packet of size B bytes arrives at time t, the following happens:

- If $Tp(t) - B > = 0$, the packet is not in excess of the peak rate and Tp is decremented by B down to the minimum value of 0, else
- The packet is in excess of the peak rate and Tp is not decremented.

According to the above definition, a positive infinite value of either PDR or PBS implies that arriving packets are never in excess of the peak rate.

Committed Data Rate Token Bucket

The committed rate of a CR-LSP is specified in terms of a token bucket C with a committed data rate (CDR). The extent by which the offered rate exceeds the committed rate can be measured in terms of another token bucket E, which also operates at rate CDR. The maximum size of the token bucket C is a committed burst size (CBS) and the maximum size of the token bucket E is EBS.

The token buckets C and E are initially (at time 0) full, i.e., the token count $Tc(0) = CBS$ and the token count $Te(0) = EBS$. Thereafter, the token counts Tc and Te are updated CDR times per second as follows:

- If Tc is less than CBS, Tc is incremented by 1, else
- If Te is less then EBS, Te is incremented by 1, else
- Neither Tc nor Te is incremented.

When a packet of size B bytes arrives at time t, the following happens:

- If $Tc(t) - B > = 0$, the packet is not in excess of the committed rate and Tc is decremented by B down to the minimum value of 0, else
- If $Te(t) - B \geq 0$, the packet is in excess of the committed rate but is not in excess of the EBS and Te is decremented by B down to the minimum value of 0, else
- The packet is in excess of both the committed rate and the EBS, and neither Tc nor Te is decremented.

Weight

The weight determines the CR-LSP's relative share of the possible excess bandwidth above its committed rate. This definition is different from our earlier definition of weighted fair queuing, and the two should not be confused. Later in this chapter, examples are provided of how weighted fair queuing can be applied to MPLS flows.

DIFFERENTIATED SERVICES, MPLS, AND TRAFFIC ENGINEERING

Chapter 8 discusses the role of MPLS in supporting differentiated services (DS). Since DS is concerned with providing QOS to network users, it must be concerned with network performance. This section explains how DS approaches traffic engineering operations.

Average Rate Meter

The Average Rate meter measures the rate at which packets are submitted to the meter over a specified time interval, for example, 1000 packets per second for a 1-second interval. If the total number of packets that arrive between the current time, T, and T-1 seconds does not exceed 1000, the packet under consideration is conforming. Otherwise, the packet is nonconforming.

Exponential Weighted Moving Average Meter

The exponential weighted moving average (EWMA) meter is expressed as follows:

avg $(n + 1)$ = $(1 - $ Gain$)$ * avg (n) + Gain * actual $(n + 1)$

t $(n + 1)$ = t (n) + Δ

where n is the number of packets, and actual (n) and avg (n) measure the number of bytes in the incoming packets in a small sampling interval, Δ.

Gain controls the frequency response of a low-pass filter operation. An arriving packet that pushes the average rate over a predefined rate, Average Rate, is nonconforming.

So, for a packet arriving at time t (m):
if (avg (m) > AverageRate)
   nonconforming
else
   conforming

Token Bucket Meter

The token bucket meter is similar to the ATM token bucket explained earlier in this chapter. Let's review the token bucket (TB) meter as defined in [BERN99].

- The TB profile contains three parameters: (a) an average rate, (b) a peak rate, and (c) a burst size.

- The meter compares packet arrival rate to average rate as byte tokens accumulate in the bucket at the average rate.
- Byte tokens accumulate in the bucket at the average rate, up to a maximum burst size (a credit).
- Arriving packets of L length are conforming if L tokens are available in the bucket at the time of packet arrival.
- Packets are allowed to reach the average rate in bursts up to the burst size, as long as they do not exceed the peak rate, at which point the bucket is drained.
- Arriving packets of L length are nonconforming if insufficient L tokens are in the bucket upon the packet arrival.

It is possible to implement token bucket models that have more than one burst size and conformance level; for example, two burst sizes and three conformance levels. This concept is known as two-level token bucket meter and is similar to Frame Relay's committed burst (Bc) and excess burst (Be) profiles.

IDEAS ON SHAPING OPERATIONS

We learned that the shapers (at the ingress LSR) condition, or shape, traffic to a certain temporal profile. For example, in an average rate meter operation in which 1,000 packets are submitted over a 1-second interval, the 1001st arrival of a packet within the 1-second interval would require the packet to be held in a buffer until it becomes conforming. Alternatively, it might be marked and possibly discarded.

Shaping operations can be complex. They must be able to prevent a "rogue" flow from seizing more QOS resources than it is allowed. Also, they must not allow conformant flows to be compromised by the rogue flows. In many instances, the shaping operations depend on the size of buffers and on queue depths. However, the shaper's individual actions are straightforward, with the use of, say, token buckets, and so on.

DS Guaranteed Rate

The Internet Network Working Group has been working on specifications to define a DS guaranteed rate per-hop behavior (PHB) [WORS98] (see Figure 7–6). The concepts revolve around non-real-time traffic with a guaranteed rate (GR). This rate is also defined in ATM as

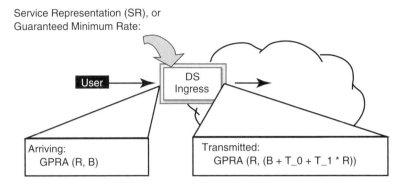

Figure 7–6   DiffServ guaranteed rate (GR).

part of the available bit rate (ABR) service. One difference between the ATM and DS approaches is that ATM is constrained to defining a successful delivery as one in which all the bits in the user frame are delivered successfully, which may entail more than one successfully delivered cell. This distinction is avoided in DS because the DS operations are defined at the L_3 IP level. The following is an overview of the DS GR, as defined in [WORS98].

The GR service provides transport of IP data with a minimum bit rate guarantee under the assumption of a given burst limit.

GR implies that if the user sends bursts of packets, which in total do not exceed the maximum burst limit, then the user should expect to see all of these packets delivered with minimal loss. GR also allows the user to send in excess of the committed rate and the associated burst limit, but the excess traffic will only be delivered within the limits of available resources.

For excess traffic, each user should have access to a fair share of available resources. The definition of fair share is network specific and is not specified by either the GR PHB or service. The DS GR uses the term *service representation* (SR) to describe a guaranteed minimum rate and the packet characteristics to which the DS GR service commitment applies. The guaranteed minimum rate uses the generic packet rate algorithm (GPRA) leaky bucket with the rate and credit parameter GPRA *(x, y)*, where *x* is the rate parameter in bytes per second and *y* is the credit limit parameter in bytes.

The SR is defined by (S, CR, BL), where S is the set of characteristics of the packet stream to which the service is being committed. The guaranteed minimum rate specification is defined as GPRA (CR, BL), where CR is the committed rate in bit/s and BL is the burst limit in bytes.

The interpretation of the SR is this: The network commits to transporting with minimal loss at least those packets belonging to the stream specified by S that pass a hypothetical implementation of the GPRA (CR, BL) located at the network's ingress interface.

The following theorem ensures a DS GR level of service that is always at least R, as defined in the GR PHB; I quote directly from Worster [WORS98]:

> Let a_j be the arrival time of the start of packet j, let t_j be the time when the start of packet j is transmitted, and let TL_j be the total length of packet j. Suppose the transmission times satisfy $s\_j < t\_j < s\_j + T\_2$, where $s\_{(j+1)} - s\_j <= (TL\_j/R)$, and also suppose that if a packet arrives when no other packets in the stream are awaiting transmission, then $a\_j + T\_0 <= s\_j < a\_j + T\_0 + T\_1$. T_0, T_1, and T_2 are, respectively, the fixed empty-queue packet latency, the maximum variation in the empty-queue packet latency, and the scheduling tolerance. Then, if the arriving packets all pass GPRA(R, B), the transmitted packets will all pass GPRA(R, $(B + (T\_0 + T\_1)*R)$).

The proof for this theorem is found in an ATM Forum paper by [WENT97].

> Though perhaps not all "guaranteed rate" nodes will schedule packets in a way that fits this form, the preceding theorem suggests that it is reasonable to expect that a significant class of such devices would have the ability to guarantee that if the input packet stream satisfies GPRA(R, B), then the output packet stream will satisfy GPRA(R, B + BTI), where BTI is the device's burst tolerance increment for the stream in question. This result allows us to consider several possible schemes by which an edge-to-edge guaranteed rate service commitment may be made. For example, if we know that each node has a BTI that does not exceed BTI_max, then we can establish GR service with parameters {S, CR, and BL} by provisioning a GR PHB with parameters {S, R = CR, and B = BL + BTI_max} along the stream's path through the network. We do not attempt to specify the rules by which a network operator should distribute appropriate GR PHB parameters. To some extent, the appropriate scheme will depend on characteristics of the implementation of the GR PBH in network nodes. It may also depend on limitations of the protocol used to distribute the parameters. GR service can also be supported across concatenated GR diff-serv networks.

Assured and Expedited Forwarding PHBs

The Internet Network Working Group has recognized that additional PHBs must be defined for DiffServ nodes to support a diverse user community. To that end, [JACO99] has authored the RFC 2598 "An

Expedited Forwarding (EF) PHB," and [HEIN99] has authored the RFC 2597 "Assured Forwarding PHB Group."

The codepoint (explained in Chapter 8) for the expedited forwarding (EF) is 101110. The DS traffic conditioning block must treat the EF PHB as the highest priority of all traffic. However, EF packets are not allowed to preempt other traffic. Consequently, a tool, such as a token bucket, must be part of the DS features. RFC 2598 includes an appendix (Appendix A of that work) that explains the results of some simulations of models to support EF PHB. I found this information very useful in my work, and I recommend you read it.

EXAMPLES OF WFQ AND MPLS FLOWS

We now provide some examples of how WFQ can be used to allocate bandwidth among different MPLS flows. Recall that WFQ assigns a weight (or priority) to each flow; it is "precedence" aware and determines the transmit order for queued packets.

We use the example cited earlier; let's review it. A SONET OC-3 link in Figure 7–7 with a 155.52 Mbit/s capacity can accept an absolute maximum of 353,207 ATM cells per second. Given the assumption that the system is consuming the OC-3 bandwidth perfectly (which depends on the efficiency of the LSR), the following holds: 155,520,000 (less the overhead of the 155.52 Mbit/s OC-3 frame) yields a rate of 149.760 Mbit/s) / 424 bits in a 53-octet cell = 353,207 cells per second.

Eight levels of priority are permitted. There are eight levels because (a) the IP TOS precedence field is 3-bits and could be used by the user application to signal to the network the user's precedence needs, and (b) the MPLS shim header Exp field is also 3-bits and can carry the TOS precedence bits. In addition, Cisco routers use this approach for native-mode IP packets.

For this example, all eight queues are to be serviced every second. Based on the weights assigned to the eight queues, n number of cells will be extracted from each queue and sent onto the SONET link. The limit is 353,207 cells during the 1-second service cycle.

A single flow is in each queue; each flow receives part of the bandwidth based on this simple scheme:

Total Weights: $8 + 7 + 6 + 5 + 4 + 3 + 2 + 1 = 36$

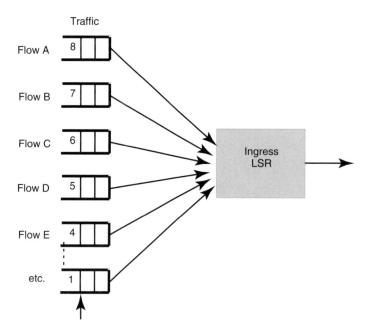

Figure 7–7 WFQ and MPLS.

The MPLS flow in the highest precedence queue is accorded 8/36 of the bandwidth. The MPLS flow in the lowest-precedent queue is accorded 1/36 of the bandwidth. Translating these functions to percentages,

$$8/36 = .222, \text{ and } 1/36 = .027$$

Consequently, flow A has 78,411 cells extracted from its queue during the service cycle, almost one-quarter of the total capacity of the link. The lowest-precedence flow has 9,536 cells serviced. Keep in mind that each flow likely consists of more than one end-user traffic flow. After all, that is one purpose of MPLS: to aggregate flows.

WFQ is more flexible than the operations shown in this example. Let's assume that multiple flows are associated with the eight traffic classes and that it is still desirable to allocate bandwidth fairly among all flows. For this example, the FEC class (and associated flow) is inferred from the MPLS label and perhaps the Exp field. Therefore, many flows can be identified. As shown in Figure 7–8, four flows are associated with precedence 5, two flows with precedence 4, and one flow with the others.

$$8 + 7 + 6 + 5(4) + 4(2) + 3 + 2 + 1 = 55$$

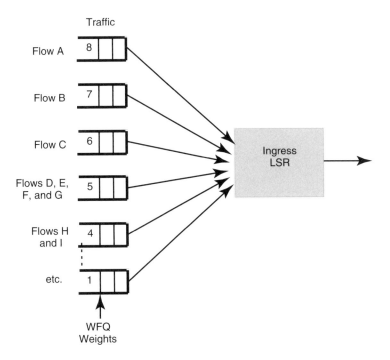

Figure 7–8   Multiple MPLS flows per traffic class.

With this set of flows, each flow in precedence 5 gets 5/55 of the bandwidth (32,109 cells per second for each flow), and each flow in precedence 4 gets 4/55 (25,687 cells per second for each flow). The one flow in queue A (Flow A) gets 8155 (51,215 cells) of the bandwidth.

With Cisco routers, WFQ allows the number of queues to be configured. Each queue corresponds to a different flow. Thus, each of the four flows with a weight of 5 in the aforementioned example would be placed into a different queue (i.e. four queues, each with a weight of 5). But each queue within this flow would still receive the same level of service.

EXAMPLES OF MPLS PROTECTION SWITCHING

MPLS supports the concept of protection switching and backup routes. An MPLS network can be set up to ensure that a link for node failure will not create a situation in which the user traffic is not delivered. There are several ways to provide protection switching. Figure 7–9 shows one approach to recover from a link failure.

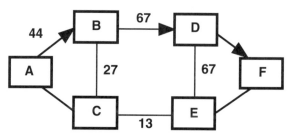

(a) Primary route (arrows)

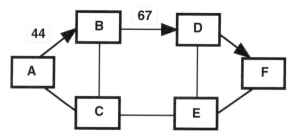

(b) Label use on primary route

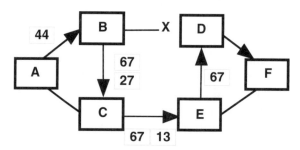

(c) Label use on secondary route

Figure 7–9   Label stacking and protection switching.

In Figure 7–9(a), the traffic is forwarded across the primary LSP from LSR A to LSR F, through LSRs B and D. The other labels shown in Figure 7–9(a) are the labels for the backup path; they are explained shortly.

As shown in Figure 7–9(b), labels 44 and 67 are used for this LSP, and at LSR D, a label pop terminates the MPLS tunnel.

In Figure 7–9(c), the link between LSR B and D fails. LSR B detects this failure (by not receiving an acknowledgment to its Hello messages from LSR D). By prior arrangement, LSR B knows that the backup path for this tunnel is to LSR C and that the label for this part of the tunnel is 27. LSR is configured to push label 67 into the stack behind label 27. Recall that label 67 was to be used at LSR D.

A label swap occurs at LSR C (27 for 13). Label 67 is not examined, since it is not at the top of the stack. At LSR E, label 13 is popped, leaving label 67 as the only label that arrives at LSR D. LSR D is configured to know that this label is associated with the same tunnel as the one with the same label number emanating from LSR B.

The example in Figure 7–9 shows a per-platform label arrangement. Node D does not associate label 67 with a specific incoming interface. The packet with label 67 arriving on the interface from node B is the primary LSP, and the packet with label 67 arriving on the interface from node E is the secondary LSP.

Figure 7–9 shows the recovery from a link failure. The failure of a node can also be recovered. This operation is shown in Figure 7–10. Since node D is down and cannot be used, node B diverts the traffic to node C by using a secondary LSP identified with label 27. At nodes C and E, labels 13 and 95, respectively, are used to forward the packet to the egress node F.

Method 1: Predefined Secondary LSP. The operations shown in Figure 7–10 can be executed in several ways. One method, arbitrarily named method 1, shown in Figure 7–11, is to premap a secondary LSP, also called a detour LSP in some literature. A generic explanation is first provided, and shortly, RSVP is brought into the picture to show how a specific protocol flow can be used to set up this method of protection switching.

It can be seen that each LSR in Figure 7–11 has an alternate route to the destination address stored in its label cross-connect table (the

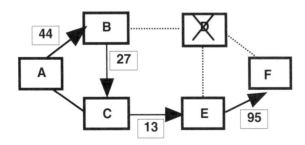

Figure 7–10   Protection switching.

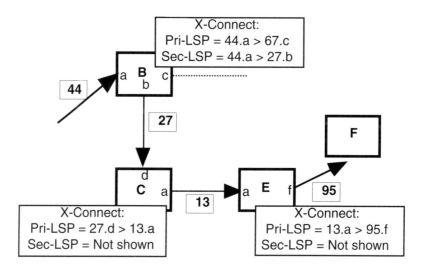

Figure 7–11   Method 1: Protection switching with predefined secondary LSP.

LFIB or NHFLE). Because the alternate routes are already set up, the recovery from a link or node failure is quite fast. Depending on the traffic load and buffer design at the routers, it might happen that no traffic is lost. If data traffic is indeed lost, the end user can use TCP to resend the packets. If the lost traffic is voice or video, the number of lost packets running over UDP will not create a severe long-term degradation of the image at the end application.

Method 2: Rebuilding the LSP.   Figure 7–12 shows another method to achieve protection switching. Essentially, nodes B, C, E, and F start from scratch and rebuild their label cross-connect tables with the exchange of label distribution (binding) messages. This method can be invoked for a restart procedure, for example, when the nodes have become unsynchronized due to problems.

Method 3: Rebuilding an LSP Segment.   Figure 7–13 shows yet another example of how protection switching can be implemented. In this scenario, node B recovers by using the liberal retention mode and rebuilds a segment of the LSP. Node B knows from its OSPF/BGP operations that the destination address is reachable by another route and that the next hop for this route is node C. It also knows that label 27 has been advertised by node C for this destination prefix. Therefore, node B

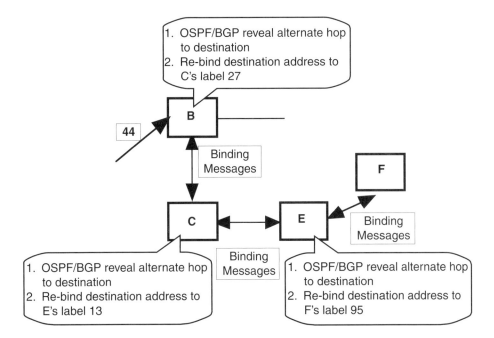

Figure 7–12   Method 2: Recreating the LSP.

replaces the primary LSP with the secondary LSP in its cross-connect table (this operation was also explained in Chapter 4, see Figure 4–8). How about the other nodes? In this example, they have already populated their cross-connect tables with the LSP. Consequently, the only change needed is that at node B.

Using RSVP to Establish Alternate/Detour LSPs

As of this writing, the operations in method 1 are being defined in [GAN01] with a process called fast-reroute. This working draft is still a bit sketchy, but it is complete enough to enable us to examine its major features.

Figure 7–14 shows how fast-reroute operates, using the topology from [GAN01]. The primary LSP runs through nodes A, B, C, D, and E. The detour LSPs use nodes F, G, H, and I. The example does not show backup operations at nodes F, G, H, and I. These nodes are most likely supporting primary LSPs for other users and perhaps using nodes A, B, C, D, and E for detour LSPs; these operations are not important to this discussion of RSVP fast-reroute.

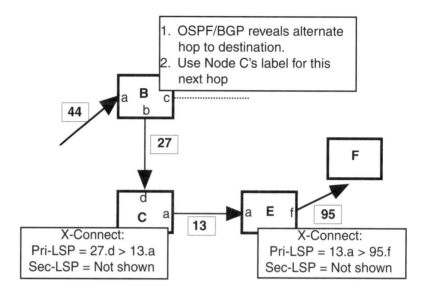

Figure 7–13   Method 3: Recreating an LSP segment.

The principal requirement for this operation is the same as in any robust network: The topology must be designed to allow the nodes to divert traffic around (a) failed links, (b) failed nodes, or (c) both (a) and (b). For example, if the traffic is at node C and the link to D or D itself fails, then node C can divert the traffic to node H.

Two new objects have been defined to support fast-reroute. They are the FAST_ROUTE and DETOUR objects. The FAST_REROUTE object is

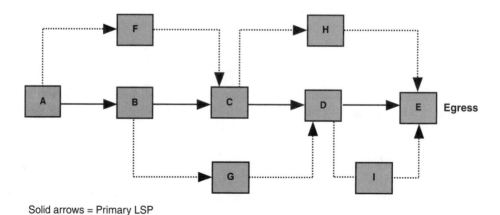

Solid arrows = Primary LSP
Dashed arrows = Detours

Figure 7–14   Network topology to demonstrate RSVP fast-reroute.

used on the main LSP, and the DETOUR object is used on the alternate (detour) LSP or LSPs. Both objects use constrained routing to specify the nodes to be either used or avoided for the LSPs. For example, in Figure 7–14, node B can send RSVP messages to its downstream nodes to dictate whether the LSP is main or detour and to specify which nodes are to be included in the LSP or excluded from the LSP. These downstream nodes can then send supporting information to their downstream nodes. The information about the downstream nodes is available through the use of the prior implementation of RSVP messages containing the RECORD_ROUTE object.

Determining Which Path Protection Method to Use

There is no "best" way to implement path protection in an MPLS network. All three methods explained above are supported in the MPLS specifications or are available in vendors' products. The method chosen depends on how fast the network must come to convergence after a failure, how big the LIB and LFIB must be to store information based on the liberal retention or conservative retention modes, and how much overhead is consumed in advertising detour LSPs (using RSVP as the example). The method chosen will be specific to the network and all are effective in providing a robust network and supporting the traffic engineering operations described in the first part of the chapter.

SUMMARY

This chapter explained three aspects of MPLS: (a) how networks in general are engineered to provide efficient services to their customers, (b) how MPLS plays a role in supporting these services, and (c) how protection switching provides robust LSPs for the end user. The importance of queue management was emphasized, and we examined several traffic engineering algorithms, with the emphasis on token buckets and WFQ.

8

OSPF in MPLS Networks

This chapter explains how extensions to OSPF allow the network user to influence how the network sets up routes in an IP routing/MPLS switching network. The user device is typically a router and is called either customer premises equipment (CPE) or a customer edge device (CE). The network node that communicates with the user node is known by several names: ingress router, ingress LSR, edge device, or customer edge device.

The reader who is familiar with OSPF can skip to the section titled "Revising OSPF to Support Constrained Routing and TE."

REVIEW OF IP-BASED ROUTING PROTOCOLS

Let's review the major aspects of the IP-based routing protocols that are pertinent to MPLS. This review must be general since the subject could translate into scores of books. If you want to delve into more details, see [BLAC00a].

To connect networks together so that they can exchange information and to move traffic through these networks efficiently, we need a method whereby a specific path (a route) is found among the many nodes (routers, servers, workstations) and routes that connect two or more network users. But it is not just a route that allows traffic to be exchanged between the users; it is the "best" route between these users.

The term is best defined according to what is considered important in the support of the user traffic. For a real-time video conference, best might be a route that offers the lowest and most consistent delay. For a funds transfer to a bank, best might be a route that offers encryption services.

Whatever best means, the identification of a route entails a route discovery operation. In its simplest terms, route discovery is the process of finding the best route between two or more nodes in an internet.

ROUTING DOMAINS

A key concept in routing is a routing domain, depicted in Figure 8–1. The routing domain is an administrative entity, and its scope depends on many factors that are eventually determined by a network administrator, such as an ISP. The term scope means how many networks and subnets are associated with the domain. A small domain consists of a few subnets; a large domain consists of many. The size of the routing domain is relative, but its goal is to establish boundaries for the dissemination of routing information; in effect, to limit the distribution of the advertisement, a concept called packet containment. If the domain contains many networks, it is likely that more routing packets must be exchanged than is the case in a domain with fewer networks.

Route Advertising

Figure 8–1 shows how the components can be configured inside the domain. For optimization of route advertising, hierarchical routing domains are created. In this example, routing domain A (RD A) is divided into two subdomains: RD A1 and RD A2. For traffic that does not pass beyond an RD boundary, it is not necessary to know (or do route advertising) about any nodes outside the domain; once again, the idea of packet containment comes into play.

The hosts and servers attached to networks 1 and 2 in RD A1 have their presence made known to each other through route advertising packets, such as OSPF advertisements. If they do not send or receive packets outside this domain (which is a common practice inside private enterprises), there is no need for other routing domains to know about them.

In most situations, a router acts as the conduit for passing end-user traffic into and out of the domain. The router also acts as the conduit for passing route advertising information between routing domains. Each router in the domain is responsible for advertising its links to the other

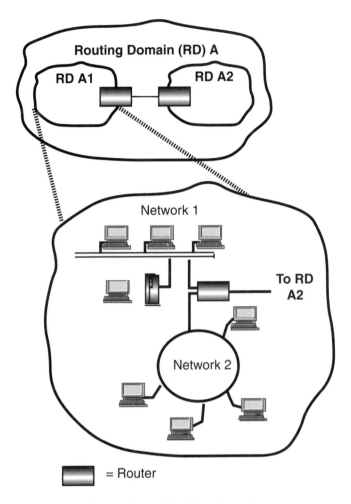

Figure 8–1   Routing domain.

routers. These advertisements are called (in OSPF) link state advertisements (LSAs). The total of all the information in these LSAs in stored in
each router in a link state database.

In many situations, a designated router is assigned the task of route
advertising for a network or a routing domain; if more than one router is
attached to the network, one of them is specified as the primary router.

Routing Domains and the "Flat Network" Problem

The information that is advertised between domains is "filtered."
The term filtered means that not every advertising packet in one domain

is sent to another domain. Instead, summary or aggregated information is given to the other domains. This idea is central to the design philosophy behind hierarchical routing domains and the idea of packet containment.

The hierarchy concept obviates a flat network topology. A flat network requires each switching node to maintain a routing table of the entire topology of a routing domain or even multiple routing domains. This approach is not feasible for large internets. Thus, the routing hierarchy concept is designed to scale well for large internets.

AUTONOMOUS SYSTEMS

Even though local authorities may administer individual networks, it is common practice for a group of networks to be administered as a whole system. This group of networks is called an *autonomous system* (AS). Examples of autonomous systems are networks located on sites such as college campuses, hospital complexes, and military installations. The networks located at these sites are connected by a router, and since these routers operate within an autonomous system, they often choose their own mechanisms for routing data.

The local administrative authorities in the autonomous systems agree on how they provide information (advertise) to each other regarding the "reachability" of the host computers inside the autonomous systems. The advertising responsibility can be given to one router, or a number of routers can participate in the operation.

How Autonomous Systems Are Numbered

The autonomous systems are identified by autonomous system numbers. How this is accomplished is up to the administrators (and the numbers are assigned by Internet administration entities), but the idea is to use different numbers to distinguish different autonomous systems. Such a numbering scheme might prove helpful if a network manager does not want to route traffic through an autonomous system which, even though it might be connected to the manager's network, may be administered by a competitor, does not have adequate or proper security services, or for other reasons. By the use of routing protocols and numbers identifying autonomous systems, the routers can determine how they reach each other and how they exchange routing information.

Autonomous systems are identified with AS numbers assigned from 1 – 65,535, with 1 – 65,411 for registered Internet numbers and 65,412 – 65,535 for private numbers.

How a Host Is Made Known to Other Domains

I just mentioned that in the Internet or in internets, it is a common approach to establish hierarchies of routing domains (levels of domains). In Figure 8–2, the two routing domains explained earlier (RD A and RD B) are connected. The two routers in these domains have been configured to be domain border routers; they are responsible for the exchange of routing information on behalf of their respective routing domains.

The hierarchy in this figure is as follows: RD A is divided into two other routing domains: RD A1 and RD A2. Likewise, RD B is divided into two other routing domains: RD B1 and RD B2. Each of these four subdomains also has a designated router (or routers) responsible for route advertising for their respective domains.

Once again, the attractive aspect of the hierarchical approach to internetworking is the practice of using routing domains to perform summary or aggregated advertising. For example, the router at RD A can use only one route advertisement packet to advertise multiple hosts and networks within the domain.

To see how, let us assume RD B is responsible for networks with addresses in the range of 192.168.1.0 through 192.168.100.0. For this example, the third decimal digit identifies specific networks (as a general practice, called subnets) 1 through 100. The fourth decimal digit identifies the hosts attached to these networks. RD B1 is responsible for subnets 192.168.1.0 through 192.168.50.0, and RD B2 is responsible for subnets 192.168.51.0 through 192.168.100.0.[1]

The RD B domain border router sends an OSPF advertisement packet (a link state advertisement, or LSA) to the RD A domain border router. The advertisement states that all IP datagrams with an IP destination address that begins with 102.168 can be sent to RD B. This information is shown in Figure 8–2 as "192.168.0.0/16."

You might know that the value of 16 is called a prefix. It serves as an address mask; the 16 means that the mask is 16 bits in length and

---

[1]A few of the IP addresses are reserved for special uses. To keep the discussion relatively simple, these examples assume the use of all numbers in an IP address range.

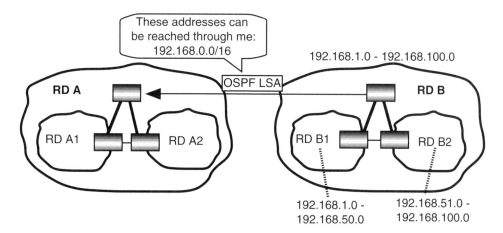

Figure 8–2   Connecting high-level routing domains (RD).

that it covers the first 16 contiguous bits of the address in front of it, namely, 192.168.

Let us assume that each of the 100 networks in RD B may have as many as 254 hosts attached (after reserved numbers are deducted). Therefore, this one advertisement serves the purpose of advertising 254 hosts x 100 networks = 25,400 addresses. Furthermore, the routing tables at the domain routers do not need to store 25,400 addresses, but only the aggregated address and its prefix. Of course, when the IP datagram reaches its final destination subnet, then the full address (with the host number) must be used to forward the traffic to the correct host.

CORRELATION (BINDING) OF THE PREFIXES TO MPLS LABELS

In Figure 8–2, the router RD B has updated its label information base (LIB) with a label associated with prefix 192.168.0.0/16. Assume the label is 76. After the routing protocol (in this example, OSPF) has done its job, the other routers that know about this prefix can also populate their label information bases.

The prefix/label binding thus far is local to each router. To bind label 76 to prefix 192.168.0.0/16, the router at RD B sends a label binding message to the router at RD A. This operation is shown in Figure 8–3.

Recall from Chapter 4 that the decision made by the RD A node to use label 76 for prefix 192.168.0.0/16 is usually based on whether the advertising node (the RD B node in this example) is the next hop in the

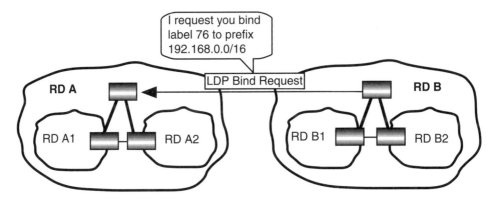

Figure 8–3   Label distribution uses the OSPF information.

downstream direction toward the destination prefix. In this example, the advertising router is the next hop in the downstream direction to this prefix, so RD A will indeed use label 76 for any traffic it sends whose IP address begins with 192.168.

MULTIPLE ROUTING PROTOCOLS

In most routing domains, more than one routing protocol is used, for several reasons.

One reason is the simple fact that the art and science of route management continues to improve, with the resultant implementation of new protocols. Yet older systems still must use the legacy protocols.

Another reason is that some of the routing protocols have been developed by vendors, others by standards groups, and there has been no clear "winner" among these systems. Therefore, it is not unusual for one network to support one type of routing protocol and another network to support a different one.

There are other reasons. Networks come in many flavors, and they have different needs. For example, route management requirements inside routing domains are often different from those protocols that operate between these domains.

In some domains, it may be important to be able to calculate routes immediately and update the routing tables quickly in the event of changes. In other domains, rapid route updates may not be important, but it may be important to have address aggregation because the number of addresses in the domain is limited.

Usually, administrative routing policies within a domain are not as important as they are between domains. Typically, within the domain, the main concern is the best route. To be sure, the best route is important between domains, but that consideration may be overridden by policy concerns, say, between two ISPs and their peering arrangements with each other. Indeed, a "best" route may not be implemented between ISPs because of administrative considerations.

As a consequence of these varying needs, routing protocols are designed to handle special needs, and, therefore, more than one approach is appropriate.

OVERVIEW OF THE ROUTING PROTOCOLS

Public and private internets have implemented a number of routing protocols, some of which have become international standards. Prevalent examples are shown in Table 8–1.

The Routing Information Protocol (RIP) was designed by the Xerox Palo Alto Research Center (PARC) for use on LANs, although it is used today in many WANs. RIP had some design flaws when it was introduced into the industry. Several have been corrected by RFCs or vendor-specific solutions.

The Open Shortest Path First (OSPF) protocol solves some of the problems found in RIP. OSPF is widely used in the industry. Its counterpart in the OSI protocol stack is the Intermediate System-to-Intermediate System protocol (IS-IS). It is not used in the Internet (and not used much elsewhere) and is not discussed in this book.

The Border Gateway Protocol (BGP) performs route advertising between the routing domains in the Internet. It overcomes many of the problems of the old EGP. BGP is a prevalent protocol in the Internet and is used between the routing domains of ISPs.

A relative newcomer to the industry is the Private Network-to-Network Interface (PNNI). It is based on using ATM in the network(s) and provides two major functions: (a) route advertising and network topology analysis, and (b) connection management (setting up and tearing down ATM connections).

Cisco implements proprietary routing protocols called the Inter-Gateway Routing Protocol (IGRP), and the Enhanced IGRP (EIGRP). EIGRP has replaced IGRP in many systems. They are similar to RIP but have several enhanced features.

Table 8–1   Routing Protocols

Routing Information Protocol (RIP)

    Intended for use on broadcast LANs

    Widely used today, with several variations

Open Shortest Path First (OSPF)

    Designed to overcome limitations of RIP and others

    Widely used today

Intermediate System to Intermediate System (IS-IS)

    Designed by Digital and part of OSI (similar to OSPF)

Border Gateway Protocol (BGP)

    Overcomes some of the limitations of EGP

    Preferred protocol between ASs

Private Network-to-Network Interface (PNNI)

    A newcomer

    Based on the use of an ATM network

Inter-Gateway Routing Protocol (IGRP) and Enhanced IGRP (EIGRP)

    Cisco's "RIP" with metric advertising and other improvements

OVERVIEW OF OSPF

OSPF protocol is an internal gateway protocol (IGP) that operates within an autonomous system. So, the OSPF routers are within one autonomous system. OSPF is a link state, or shortest path first protocol, in contrast to some of the earlier Internet protocols that are based on some type of fewest number of hops approach. The protocol is tailored specifically for an internet and includes such capabilities as subnet addressing and type of service (TOS) routing.[2]

OSPF bases its route discovery decisions on addresses and link state metrics. OSPF is an adaptive protocol in that it adjusts to problems in the network (a link or node failure) and provides short convergence periods to stabilize the routing tables. It is also designed to prevent looping of traffic, which is quite important in mesh networks or in LANs where multiple routers may be available to connect different LANs.

---

[2] TOS routing is not deployed extensively in the Internet, although the intent in the original specifications was that it be an important aspect of OSPF. Check your installation for possible TOS support.

OSPF is encapsulated into the IP datagram data field. The IP protocol ID for OSPF is 89.

Role of the Router in OSPF

OSPF permits a router to assume several different roles in an OSPF routing domain. It can act as a designated router for an autonomous system, a designated router for an area within an autonomous system, and a designated router for a network to which multiple routers are attached. Within these routing domains, the router may send and receive several different types of link state advertisements (LSAs). Some LSAs are for handshakes, such as a hello packet, between the routers; others contain information about a node's database; still others are update packets.

Directed Graphs

OSPF works with directed graphs, as shown in Figure 8–4(a). The graphs contain values between two points, the interfaces between two routers (their link interfaces). The values represent the weighted shortest path value with the router established as the root. Consequently, the shortest path tree from the router to any point in an internet is determined by the router performing the calculation. The calculation only determines the next hop to the destination in the hop-to-hop forwarding process. The link state database used in the calculation is derived from

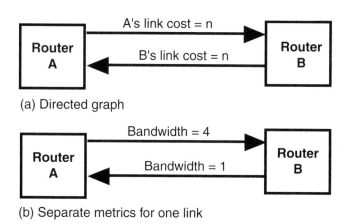

(a) Directed graph

(b) Separate metrics for one link

Figure 8–4   Directed graphs.

the information obtained by advertisements sent by the routers to their neighbors, with periodic flooding throughout the routing domain.

The information focuses on the topology of the network(s) with a directed graph. Routers and networks form the vertices of the graph. Periodically, this information can be broadcast (flooded) to all routers in the autonomous system or area, or sent as needed after a change. An OSPF router computes the shortest path to the other routers in the routing domain from information in the link state database. If the calculations reveal that two paths are of equal value, OSPF will distribute the traffic equally on these paths.

Conceptually, separate cost metrics can be computed for both directions on a link, as shown in Figure 8–4(b). Moreover, most implementations simply use the same value for each direction on the link.

Key Operations

OSPF implements a hello protocol. It is a handshaking protocol that enables the routers to learn about each other, exchange information, and later perform pings with neighbor routers to make certain the link or router is up.

After the hello operations have been completed, the peer routers are considered to be *merely adjacent*. This term means the routers have completed part of the synchronization, but not all of it.

Next, the routers exchange information that describes their knowledge of the routing domain. This information, called a database description, is placed in link state advertisement (LSA) messages. The database descriptions are not the entire link state database but contain sufficient information for the receiving router to know if its link state database is consistent with its peer's databases. If all is consistent, the neighbor is now defined as *fully adjacent*. Otherwise, the routers then exchange LSAs containing link state updates, eventually becoming fully adjacent.

Thereafter, periodic hellos are issued to keep peers aware of each other. Also, the LSAs that the router originated must be sent to its peers every 30 minutes to make certain all link state databases are the same.

OSPF is concerned with link state database synchronization, and much of the OSPF code is devoted to this very task.

OSPF Areas

Enterprises with large systems may operate with many networks, routers, and host computers. For management of this vast array of communications components, it is quite possible that many LSAs must be exchanged between the routers to determine how to relay traffic within the autonomous system between the sending computer and the receiving computer.

The network administrator must evaluate how much routing traffic is to be sent between the routers because this routing traffic can affect the throughput of user data. And we know that a common practice in route advertising and route discovery is to flood the advertisements to all nodes in the routing domain. Although measures can be taken to reduce the amount of duplicate traffic that a node receives, it is not unusual for networks to have their routing nodes connected in such a fashion as to create loops, which means that it is possible for an advertisement to be received more than once. OSPF does a good job of packet containment and managing loops.

One approach that is used by OSPF is to divide or partition the autonomous system into smaller parts, called *areas*. This approach reduces the amount of routing traffic that is sent through the autonomous system because the areas are isolated from one another. This practice reduces the amount of information a router must maintain about the full autonomous system. Also, it means that the overhead information transmitted between routers to maintain OSPF routing tables is substantially reduced.

A designated router, say, router 3 in Figure 8–5, assumes the responsibility for informing the routers in the area about the other routers, networks, and hosts residing in the autonomous system.

Packet Containment

As a consequence of this approach, the routers within the area are not concerned about the details of the full autonomous system (the other areas). They obtain their information from a designated router, in the example in Figure 8–5, router 3.

Assuming network 3 is a broadcast network, such as an Ethernet, OSPF uses multicasting to restrict LSA packet processing at nodes that do not need to examine certain routing packets. Assuming network 3 is a nonbroadcast network (a switched network, like Frame Relay), OSPF uses a packet "filtering" procedure to reduce the number of routing packets that are exchanged between the routers in the area.

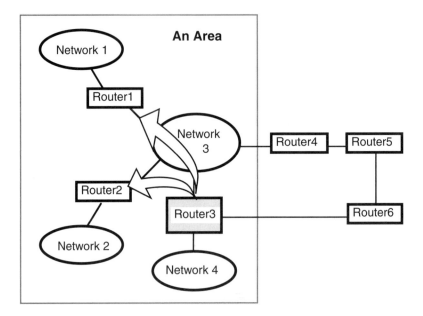

Figure 8–5   Route advertising within the area.

Stub Areas

In this example, networks 1, 2, and 4 are stub networks. OSPF also supports the concept of *stub areas*. The stub area is one into which routing information on external routes is not sent. Instead, the area border router generates a default route for destinations outside the area, and the routers in this stub area use this route. After all, why have all this routing information going about, when the stub area really does not act on it anyway?

In addition, the network manager can set up the area border router to prevent it from sending summary link advertisements into the stub area. These summary link advertisements are designated as type 3 LSAs.

REVISING OSPF TO SUPPORT CONSTRAINED ROUTING AND TE

The idea of the OSPF revisions to support MPLS TE is simple: more attributes are added to the OSPF link advertisements. In this part of the chapter, these attribute extensions are examined in light of [KATZ01], [KOMP02], [MANN01], and my thoughts about the subject.

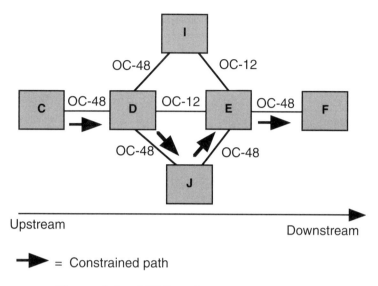

Upstream

Downstream

= Constrained path

Figure 8–6   OSPF and constrained routing.

GENERAL IDEA OF USING OSPF IN MPLS NETWORKS

OSPF is so named because it selects a "shortest path" from a source node to a destination node. A better term is optimum path, but the former term is the prevalent one used. OSPF can be used in MPLS networks to assist in building a constrained path for an LSP, in accordance with TE requirements for the LSP.

To illustrate, Figure 8–6 is a view of the core LSRs in our network model. The LSP from node C to Node F is to be based on a minimum bandwidth of OC-48; that is, the LSP is constrained to support at least a bandwidth of OC-48. To establish a constrained TE LSP for this requirement, the following events take place (more events occur within the OSPF interactions; this example summarizes them):

1. Event 1: Node C adds its adjacent nodes (node D) to the shortest path candidate list. Obviously, the C-D link meets the OC-48 bandwidth requirement.
2. Event 2. The neighbors of node D—nodes I, E, and J—are examined. Links D-I and D-J also meet the minimum bandwidth requirements, and therefore the links do not violate the constraints imposed on the LSP. LSRs I and J are added to the OSPF candidate list for this LSP. Node E is not added to this list because the D-E link does not meet the bandwidth requirements of the LSP.

3. Event 3: Since both LSRs I and J exhibit the same metric to LSR D, one is chosen for the next iteration of the OSPF algorithm. Let's choose LSR I (because of its address value, for example). It does not make any difference which node is chosen.

4. Event 4: The links of LSR I to its neighbors are examined. It is found that the link to LSR E does not meet the requirements for the LSP. Therefore, LSR E (once again) is not added to the candidate list for this LSP.

5. Event 5: LSR J is still in the candidate list, so its links are examined. It is found that the D-J link (known earlier) and the J-E link meet the LSP's bandwidth requirements. Consequently, LSR J is added to the OSPF candidate list.

6. Event 6: LSR J's neighbors are examined, revealing LSR E. The E-F link to LSR F meets the requirement, so LSR F is now placed in the candidate list.

7. Event 7: It is known that LSR F is the destination node for the address being sought, so the algorithm is stopped. The constrained LSP for the example in Figure 8–6 is C-D-J-E-F

With these ideas in mind, we now examine the efforts of the IETF Network Working Group in defining extensions to OSPF to support traffic engineering in an MPLS routing domain [KATZ01].

TRAFFIC ENGINEERING EXTENSIONS TO OSPF

[KATZ01] defines OSPF "extended link attributes," which are really nothing more than the addition of more attributes to links in OSPF advertisements. The idea is to use the extensions to build an "extended" link state database just as router link state advertisements (LSAs) are used to build a "regular" link state database. The difference is that the link state database (called the traffic engineering database) has additional link attributes to support traffic engineering and constraint-based source routing and to monitor the extended link attributes.

The LSA TE Extension

The principal aspect of the OSPF extension for MPLS traffic engineering is the addition of the traffic engineering LSA, shown in Figure 8–7. These data are placed in an OSPF packet that is preceded by the

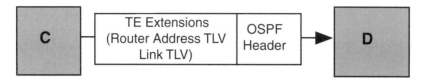

Figure 8–7   OSPF LSA TE extension.

standard OSPF header (not shown in Figure 8–7). The contents of the packet are formatted by the conventional type-length-value syntax, used in many Internet protocols. This convention is explained in Chapter 5. To review, take a look at "LDP Messages" in Chapter 5.

The TLVs

The packet contains one of two top-level TLVs, and as many as nine TLVs (called sub-TLVs) follow. This part of the chapter examines these TLVs, to help in understanding how the OSPF extension is used in MPLS traffic engineering operations.

Router Address TLV.   The Router Address TLV specifies a stable IP address of the advertising router that is always reachable if there is any connectivity to it. This value is the familiar "router ID," used in other protocols.

Link TLV.   The Link TLV describes a single link. It consists of a set of sub-TLVs.

These sub-TLVs are defined below:

1. Link type: This type describes the link as either point-to-point or multiaccess (shared bus, ring, etc.).
2. Link ID: This type identifies the other end of the link. For point-to-point links, this field is the router ID of the neighbor. For multiaccess links, this field is the interface address of the designated router.
3. Local interface IP address: This type specifies the IP address(es) of the interface corresponding to this link.
4. Remote interface IP address: This type specifies the IP address(es) of the neighbor's interface corresponding to this link. This and the local address are used to discern multiple parallel links between systems.

5. Traffic engineering metric: This type specifies the TE metric for the link. It can be different from the OSPF metric for that link.

6. Maximum bandwidth: This type specifies the maximum bandwidth (actual link capacity) that can be used on this link in the direction of the LSA advertisement, that is, from the originating node to the receiving node.

7. Maximum reservable bandwidth: This type specifies the maximum bandwidth that can be reserved on this link in this direction. Note that this can be greater than the maximum bandwidth (in which case the link may be oversubscribed).

8. Unreserved bandwidth: This type specifies the amount of bandwidth not yet reserved at each of eight priority levels. The values correspond to the bandwidth that can be reserved with a setup priority of 0 through 7, arranged in increasing order with priority 0 occurring at the start of the sub-TLV, and priority 7 at the end of the sub-TLV. Each value will be less than or equal to the maximum reservable bandwidth.

9. Resource class/color: This type specifies the administrative group membership for this link, based on a bit map that represents a resource class (or color). The idea of a color is explained in Chapter 5, see "Colored Threads," but keep in mind that there are more ways than one to implement colored threads, so check your specific implemtation of these ideas.

ROLE OF OSPF IN VPNS

VPNs are the subject of Chapter 11, and the role of OSPF in VPNs is discussed in Chapter 11. To look ahead, see the section in Chapter 11 titled "Role of OSPF in VPNs."

SUMMARY

The use of OSPF in MPLS networks is a relatively new endeavor, spearheaded by the IETF Working Groups. A few extensions to an LSA permit OSPF to assume the role of a constrained routing protocol, although the preference thus far is for the use of LDP and/or CR-LDP.

9

Constraint-Based Routing with CR-LDP

This chapter explains a traffic engineering function called constraint-based routing. The primary focus is on how LSPs are established with LDP, but other alternatives such as a modified OSPF and RSVP are discussed. Earlier chapters explained some of the basic and MPLS-oriented features of RSVP, and you may want to review that material before reading the section on RSVP.

The information in this chapter is based on the MPLS working drafts, and especially on [JAMO01], and several papers from [FORO00]. Be aware that this information is closely tied to the traffic engineering discussions in Chapter 7, which is prerequisite reading. In addition, constraint-based routing is explained in relation to ATM, Frame Relay, and RSVP in Appendix B.

THE BASIC CONCEPT

Constraint-based routing (CR) is a mechanism by which traffic engineering requirements for MPLS networks can be met. The basic concept is to extend LDP for support of constraint-based routed label switched paths (CR-LSPs) by defining mechanisms and additional type-length-values (TLVs) for support of CR-LSPs or to use existing protocols to support constraint-based routing.

211

CR can be set up as an end-to-end operation; that is, from the ingress CR-LSR to the egress CR-LSR. The idea is for the ingress CR-LSR to initiate CR and for all affected nodes to be able to reserve resources using LDP.

The term *constraint* implies that in a network and for each set of nodes there exists a set of constraints that must be satisfied for the link or links between two nodes. An example of a constraint is a path that has a minimum amount of bandwidth. Another example is a path that is secure. The protocol that finds such paths (such as a modified OSPF) is constrained to advertise (and find) paths in the routing domain that satisfy these kinds of constraints.

In addition, constraint-based routing attempts to meet a set of constraints and, at the same time, optimize some scalar metric [DAVI00]. One important scalar metric is hop count for delay-sensitive traffic. Experience has shown that extra hops create jitter, especially if the Internet is busy and the routers are processing a lot of traffic.

EXPLICIT ROUTING

Explicit routing (ER) is integral to constraint-based routing. This route is set up at the edge of the network, in accordance with QOS criteria and routing information. Figure 9–1 shows an example of explicit routing.

The explicit route starts at ingress router A and traverses B, then D, and exits at egress router F. The explicit route is not allowed to traverse

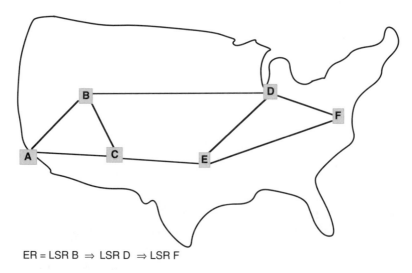

ER = LSR B ⇒ LSR D ⇒ LSR F

Figure 9–1   Explicit routing.

LSRs C and E. The allowed route can be established by means of LDP messages. Based on [JAMO01], the explicit route is coded in a label request message. The message contains a list of nodes (or group of nodes) that define the CR route. After the CR-LSP has been established, all of a subset of the nodes in a group can be traversed by the LSP.

The capability to specify groups of nodes, of which a subset will be traversed by the CR-LSP, allows the system a significant amount of local flexibility in fulfilling a request for a constraint-based route. Moreover, constraint-based routing requires the path to be calculated by the source of the LSP traffic.

LDP and Constraint-based Routing

If LDP is used for constraint-based routing, the constraint-based route is encoded as a series of ER hops contained in a CR-LDP message, explained later in this chapter. Each ER hop can identify a group of nodes in the constraint-based route, and TLVs that describe traffic parameters, such as peak rate and committed rate, are available. A constraint-based route is, then, a path that includes all of the identified groups of nodes in the order in which they appear in the TLV.

Strict and Loose Explicit Routes

An explicit route is represented in a Label Request Message (explained shortly) as a list of nodes or groups of nodes along the constraint-based route. When the CR-LSP is established, all or a subset of the nodes in a group can be traversed by the LSP. Certain operations to be performed along the path can also be encoded in the constraint-based route.

PREEMPTION

CR-LDP conveys the resources required by a path on each hop of the route. If a route with sufficient resources cannot be found, existing paths can be rerouted to reallocate resources to the new path. For example, in Figure 9–1, if node B is down or has insufficient resources to meet the QOS requirements of the FEC, node C can be selected instead.

This idea is called *path preemption*. Setup and holding priorities rank existing paths (holding priority) and the new path (setup priority) to determine if the new path can preempt an existing path.

The setup priority of a new CR-LSP and the holding priority attributes of the existing CR-LSP specify priorities. Signaling a higher holding priority expresses that the path, once it has been established, should have a lower chance of being preempted. Signaling a higher setup priority expresses the expectation that, in the case in which resources are unavailable, the path is more likely to preempt other paths. The setup and holding priority values range from 0 to 7. The value 0 is the priority assigned to the most important path. It is referred to as the highest priority. The value 7 is the priority for the least important path.

The setup priority of a CR-LSP should not be higher (numerically less) than its holding priority, since it might bump an LSP and be bumped by the next equivalent request.

CR MESSAGES AND TLVS

This part of the chapter describes the CR messages and TLVs. The contents and general purpose of each message is explained, and in subsequent sections of the chapter, more details are provided.

Label Request Message

The Label Request message is modified from LDP, and shown in Figure 9–2. Several fields in this message have been explained in Chapter 5. The new TLVs are examined in the next section of this chapter.

0	1–14	1 5	1 6	17–30	3 1
0	Label Request (0x0401)			Message Length	
	Message ID				
	FEC TLV				
	LSPID TLV (CR-LDP, mandatory)				
	ER-TLV (CR-LDP, optional)				
	Traffic TLV (CR-LDP, optional)				
	Pinning TLV (CR-LDP, optional)				
	Resource Class TLV (CR-LDP, optional)				
	Preemption TLV (CR-LDP, optional)				

Figure 9–2   Label request message.

0	1–14	1 5	1 6	17–30	3 1
0	Label Mapping (0x0400)			Message Length	
		Message ID			
		FEC TLV			
		Label TLV			
		Label Request Message ID TLV			
		LSPID TLV (CR-LDP, optional)			
		Traffic TLV (CR-LDP, optional)			

Figure 9–3   Label mapping message.

Label Mapping Message

The Label Mapping message is shown in Figure 9–3. This message is sent by a downstream LSR to an upstream LSR if one of the following conditions has been satisfied: (a) the LSR is the egress end of the CR-LSP and an upstream mapping has been requested or (b) the LSR received a mapping from its downstream next-hop LSR for a CR-LSR for which an upstream request is still pending.

Notification Message

Notification messages, shown in Figure 9–4, carry status TLVs to specify events being signaled. Notification messages are forwarded toward the LSR originating the label request at each hop.

0	1–14	1 5	1 6	17–30	3 1
0	Notification (0x0001)			Message Length	
		Message ID			
		Status TLV			
		Optional Parameters			

Figure 9–4   Notification message.

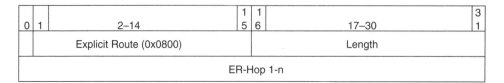

Figure 9–5   Explicit route TLV.

Explicit Route TLV

The explicit route TLV, depicted in Figure 9–5, specifies the path to be taken by the LSP that is being established. It contains one or more explicit hop TLVs, explained next.

Explicit Route Hop TLV

The explicit route hop TLV contains the hop identifiers. The TLV is shown in Figure 9–6. The ER-Hop-Type field conveys information about the contents field, indicating that this field is IPv4 prefix, IPv6 prefix, autonomous system (AS) number, or an LSPID. The L bit indicates whether the node or group operates as loose or strict routing. The content field contains the prefixes or whatever is being conveyed in this TLV.

Traffic Parameters TLV

Figure 9–7 shows the format for the traffic parameter TLV. As the name implies, it conveys traffic parameters to other CR-LSR nodes. Let's examine the fields in the TLV first, then review their functions (for continuity, these functions are also explained in Chapter 7 under "Constraint-Based Routing (CR)."

Functions of the TLV Fields.   The fields in the TLV perform the following functions.

The flag field performs a number of functions. The first two bits are reserved for future use. The remaining six bits are defined as follows. Each

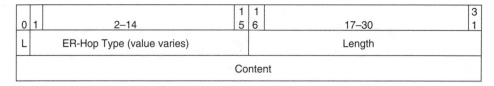

Figure 9–6   Explicit route hop TLV.

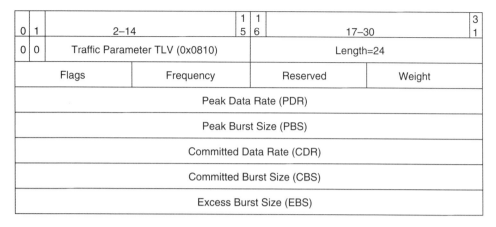

Figure 9–7   Traffic parameters TLV.

flag bit is a negotiable flag corresponding to a traffic parameter. The negotiable flag value 0 denotes not negotiable, and value 1 denotes negotiable.

F1 — Corresponds to the PDR.
F2 — Corresponds to the PBS.
F3 — Corresponds to the CDR.
F4 — Corresponds to the CBS.
F5 — Corresponds to the EBS.
F6 — Corresponds to the weight.

The frequency field is coded with the following code points defined:

0 — Unspecified
1 — Frequent
2 — Very frequent
3–255 — Reserved

The weight field indicates the weight of the CR-LSP. Valid weight values are from 1 to 255. The value 0 means that weight is not applicable for the CR-LSP.

Each traffic parameter is encoded as a 32-bit number. The values PDR and CDR are in units of bytes-per-second. The values PBS, CBS, and EBS are in units of bytes.

Functions of the Traffic Parameters. The frequency specifies the granularity of the CDR allocated to the CR-LSP. The value "very frequent" means that the available rate should average at least the CDR when measured over any time interval equal to or longer than the shortest packet time at the CDR. The value "frequent" means that the available rate should average at least the CDR when measured over any time interval equal to or longer than a small number of shortest packet times at the CDR. The value "unspecified" means that the CDR can be provided at any granularity.

The peak rate defines the maximum rate at which traffic should be sent to the CR-LSP. The peak rate is useful for resource allocation. If resource allocation within the MPLS domain depends on the peak rate value, then it should be enforced at the ingress to the MPLS domain.

The committed rate defines the rate that the MPLS domain commits to be available to the CR-LSP.

The excess burst size can be used at the edge of an MPLS domain for traffic conditioning. It can be used to measure the extent by which the traffic sent on a CR-LSP exceeds the committed rate.

Once again, to learn how these parameters are used together, see "Constraint-Based Routing (CR)" in Chapter 7. For this discussion, we continue to examine the CR-LDP TLVs.

Preemption TLV

The preemption TLV is shown in Figure 9–8. The two key fields in the TLV are SetPrio (SetupPriority) and HoldPrio (HoldingPriority). A SetupPriority of 0 is the priority assigned to the most important path. It is referred to as the highest priority. A priority of 7 is assigned to the least important path.

A HoldingPriority of 0 is the priority assigned to the most important path. It is referred to as the highest priority. A priority of 7 is assigned to the least important path.

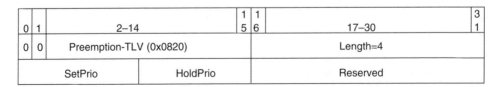

0	1	2–14	1 5	1 6	17–30	3 1
0	0	Preemption-TLV (0x0820)			Length=4	
		SetPrio		HoldPrio	Reserved	

Figure 9–8   Preemption TLV.

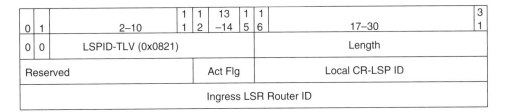

Figure 9–9   LSPID TLV.

LSPID TLV

The LSPID TLV is shown in Figure 9–9. It is a unique ID of a CR-LSP. The LSPID is composed of the ingress LSR Router ID (or any of its own Ipv4 addresses) and a locally unique CR-LSP ID to that LSR. The LSPID is useful in network management, in CR-LSP repair, and in the use of an already established CR-LSP as a hop in an ER-TLV.

An action indicator flag is carried in the LSPID TLV; the flag indicates explicitly the action that should be taken if the LSP already exists on the LSR receiving the message.

After a CR-LSP is set up, its bandwidth reservation may need to be changed by the network provider, because of the new requirements for the traffic carried on that CR-LSP. The action indicator flag (Act Flg) indicates the need to modify the bandwidth and possibly other parameters of an established CR-LSP without service interruption.

The Local LSPID identifies the CR-LSP that is locally unique within the ingress LSR originating the CR-LSP. The ingress LSR Router ID simply identifies the ingress LSR router. This TLV performs the functions of colored threads, a subject explained in Chapter 5 (see the section titled "Colored Threads").

Resource Class TLV

The resource class TLV, shown in Figure 9–10, specifies which links are acceptable by this CR-LSP and prunes the topology of the network. The RsCLs field (resource class bit mask) indicates which of the 32 nodes the CR-LSP can traverse.

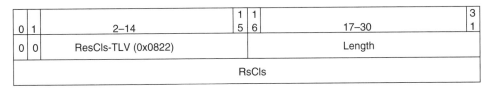

Figure 9–10   Resource class TLV.

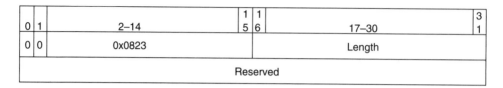

Figure 9–11   Route pinning TLV.

Route Pinning TLV

Figure 9–11 shows the TLV format for route pinning. Route pinning is applicable to segments of an LSP that are loosely routed—i.e., those segments that are specified with a next hop with the L bit set or where the next hop is an abstract node. A CR-LSP can be set up by route pinning if it is undesirable to change the path used by an LSP even when a better next hop becomes available at some LSR along the loosely routed portion of the LSP.

The P bit is set to 1 to indicate that route pinning is requested. The P bit is set to 0 to indicate that route pinning is not requested.

CR-LSP FEC TLV

A new FEC element, shown in Figure 9–12, supports CR-LSPs. It does not preclude the use of other FEC elements (Type = 0x01, 0x02, 0x03) defined in the LDP specification in CR-LDP messages. The CR-LDP FEC element is an opaque FEC to be used only in messages of CR-LSPs.

SUMMARY

This chapter explained a traffic engineering function called constraint-based routing. The primary focus was on how LSPs are established with LDP. The basic idea of constraint-based routing is to meet the traffic engineering requirements for MPLS networks. Indeed, constraint-based routing is closely associated with the traffic engineering concepts explained in Chapter 7.

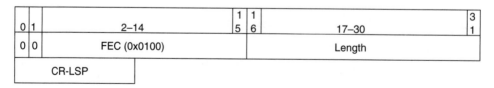

Figure 9–12   CR-LSP FEC TLV.

10

MPLS, Optical Networks, and GMPLS

Label switching and MPLS are considered by many to be key components in the new optical transport networks. Therefore, this chapter is devoted to these subjects and how they fit into optical networks. The first topic for discussion is a short tutorial on optical wavelengths, followed by a discussion of the relationships of optical cross-connect nodes to MPLS. Next, we examine of some of the major issues in internetworking MPLS with optical networks. This chapter concludes with a discussion on Generalized MPLS (GMPLS).

This chapter covers the major aspects of a wide-ranging subject. For more details on MPLS and optical networks, please refer to a companion book in this series [BLAC02].

WDM AND OPTICAL NETWORKS

Wave division multiplexing (WDM) is based on a well-known concept called frequency division multiplexing, or FDM. As seen in Figure 10–1, with this technology the bandwidth of a channel (its frequency domain) is divided into multiple channels, and each channel occupies a part of the larger frequency spectrum. In WDM networks, each channel is called a *wavelength*. This name is used because each channel operates at a different frequency and a different optical wavelength (and the higher the

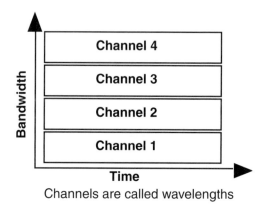

Channels are called wavelengths

Figure 10–1   WDM.

frequency, the shorter the signal's wavelength). A common shorthand notation for wavelength is the Greek symbol lambda, shown as λ.

The wavelengths on the fiber are separated by unused spectrum. This practice keeps the wavelengths separated from each other and helps prevent their interfering with each other. This idea is called channel spacing or, simply, spacing. It is similar to the idea of guardbands used in electrical systems. In Figure 10–1, the small gaps between each channel represent the spacing.

Relationships of Optical and MPLS Operations

A convenient way to view the relationship of MPLS and optical networking is through the layered model, as shown in Figure 10–2. The optical operations occur in layer 1; the MPLS operations occur in a combination of layers 2 and 3.[1]

The data plane of an LSR uses label swapping to transfer a labeled packet from an input port to an output port. The data plane of an optical switch uses a switching matrix to connect an optical switched path (OSP) from an input port to an output port. The OSP extends from a node's output interface to an adjacent node's input interface. The OSP is also called an optical trail.

Traffic from the user application at an upper layer can be sent to either the data or control plane of the MPLS lower layer on the transmit side, with a reverse operation occurring on the receive side. In fact, such

---

[1] Some of the new protocols' operations span layers 2 and 3. ATM is one example; MPLS is cited in some literature as another. Strictly speaking, MPLS is a layer 3 protocol, in that it does not define the critical function of layer 2 frame delineation.

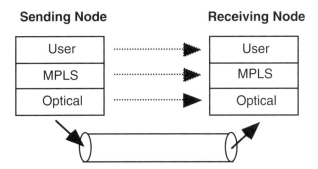

Figure 10–2   Multiprotocol lambda switching.

an approach is common. For example, an MPLS router may send a control message to an adjacent MPLS router that sets up timers for label management operations. This control message could go over either an optical data channel or an optical control channel.

An LSR performs label switching by first establishing a relation between an input port and an input label, and an output port and an output label. Likewise, an optical switch provisions an optical channel by first establishing a relation between an input port and an input optical channel (wavelength on a fiber or a fiber itself) and an output port and an output optical channel. In the LSR, the next hop label forwarding entry (NHLFE) of the LFIB maintains these input-output relations.

In the optical switch, the switch controller configures the internal interconnection fabric (called an OSP cross-connect table or a wavelength forwarding information base [WFIB]) to establish the relationships between MPLS and the optical channels.

We know from previous discussions that the MPLS control planes include resource discovery, connection management, and binding operations.

The control plane of the optical node discovers, distributes, and maintains relevant state information associated with the OSPs and establishes and maintains these OSPs under various optical internetworking traffic engineering rules and policies.

A significant difference between current LSRs and optical switches is that with LSRs the forwarding information is carried explicitly as part of the labels appended to data packets, whereas with optical switches the switching information is implied from the wavelength or optical fiber. To re-emphasize, the label is used by the LSP cross-connect table, and the wavelength is used by the OSP cross-connect table.

MULTIPROTOCOL LAMBDA SWITCHING (MPλS)

The framework for interworking optical networks and MPLS is called MPλS, and is defined in [AWDU01]. As you might expect, MPLS and optical networks have control mechanisms (control planes) to manage the user traffic. These control planes are shown in Figure 10–3. As explained often in this book, the MPLS control plane is concerned with label distribution and binding an end-to-end LSP. The optical control plane is concerned with setting up optical channels (wavelengths), optical coding schemes pertinent to the specific optical implementation, transfer rates (in bit/s), and protection switching options. Once again, the channels are called optical switched paths (OSPs), or optical trails.

MPLS and Optical Wavelength Mapping

A key aspect of MPLS and optical network interworking is to map (correlate) an MPLS label value to an optical wavelength. In Figure 10–3, the information from the LFIB (and, of course, from prior operations of the IP control plane) is used to map a label onto a wavelength representing an OSP. Next, a control protocol such as GMPLS or the Link Management Protocol (LMP) can be used by the node to inform its neighbors about the mapping of the specific label to a specific wavelength.

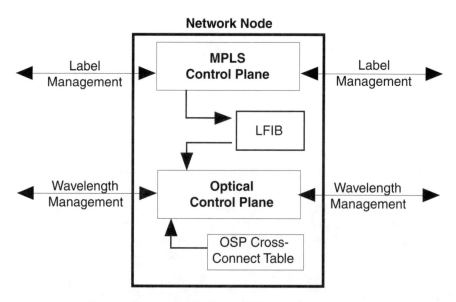

Figure 10–3   MPLS and optical control planes.

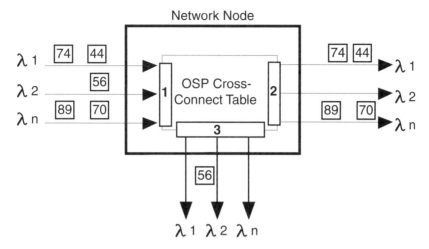

Figure 10–4   Optical O/O/O data plane.

Thereafter, the optical node need not be concerned with processing the MPLS labels. As shown in Figure 10–4, the optical cross-connect is concerned with switching (actually, reflecting) the wavelengths from the input interface to the output interface. This illustration is an example of an optical data plane that is an O/O/O device, The three O's have the following meanings: (a) the first O signifies that the incoming signal is optical, (b) the second O signifies that the data plane is all optical (with no electronic conversion of the signal), and (c) the third O signifies that the output signal is optical.

In Figure 10–4, λ2 arriving on interface 1 is switched to interface 3. The other lambdas are cross-connected to interface 2. The labels and the user payload is not examined by the switch. If labels are examined, the switch must resort to O/E/O operations: convert the optical signal to an electrical signal in order to process the label header.

Ideally, the label-to-wavelength mapping takes place at the edge of the optical network. In Figure 10–5, the user sends traffic to the network. At the ingress node (now noted as an LSR/OXC, where OXC means optical cross-connect) the MPLS label is correlated to an appropriate wavelength; that is, an appropriate channel into the network and out of the network to reach the destination user. The transit nodes, now labeled as transit OXCs, have been configured to process the wavelength to make the routing decisions.

The MPLS label may or may not be examined at the transit nodes. For this example, it is assumed the user payload is to be sent through the

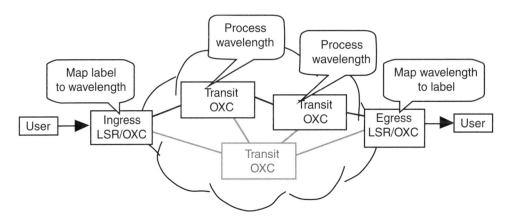

Figure 10–5   Processing the user traffic.

network on a predefined constrained LSP from the ingress LSR/OXC to the egress LSR/OXC. Thus, it is not necessary to know about the MPLS label as long as all nodes know the relationship of the wavelength associated with the label and its final destination.

An explicit LSP is one whose route is defined at its origination node or by a control protocol such as OSPF (that discovers and sets up a path through the network). However the path is defined, once it is set up, it remains stable, unless problems occur at a node or on an optical trail. Explicit LSPs and OSPs exhibit certain commonalties. They are both unidirectional, point-to-point relationships. An explicit LSP provides a packet forwarding path (traffic trunk) between an ingress LSR and an egress LSR. An OSP provides an optical channel between two endpoints for the transport of user traffic.

The payload carried by both LSPs and OSPs are transparent to intermediate nodes along their respective paths. Both LSPs and optical paths can be configured to stipulate their performance and protection requirements.

Failure of the Optical Connection

In the event of a fiber failure or a fiber node failure, there must be a method to find a backup route. In an MPλS network, there must be very close coordination between the optical and label control planes if they are indeed different software processes in the nodes. For example, if an optical connection is lost, the optical control plane must be able to inform the MPλS control plane so that neighbor LSRs can be informed of the problem. As the MPLS and the optical layers mature, it is likely that this coordination will be aided by a data link layer specially designed for this purpose.

VIEWS ON THE MPLS CONTROL PLANE
AND THE OPTICAL SWITCH

To conclude this part of the discussion of MPLS and optical networks, this section highlights some efforts taking place in the IETF regarding the subject of traffic engineering and the use of an MPLS control plane in an optical switch.

The components of the MPLS traffic engineering control plane model include the following modules [AWDU01]:

- Resource discovery: Negotiates and reserves bandwidth.
- State information dissemination: Distributes relevant information concerning the state of the network, including topology and resource availability information. In the MPLS context, distribution is accomplished by extension of conventional IP link state interior gateway protocols to carry additional information in their link state advertisements.
- Path selection: Selects an appropriate route through the MPLS network for explicit routing. It is implemented by constraint-based routing.
- Path management: Manages label distribution, path placement, path maintenance, and path revocation.

Control Adaptation

In adapting the MPLS traffic engineering control plane model to OXCs, a number of issues should be considered. One issue concerns the development of optical-specific domain models that abstract and capture relevant characteristics of the optical/MPLS network. The domain models help to delineate the design space for the control plane problem in OXCs and to construct domain-specific software reference architectures.

Two domain models have been identified: (a) a horizontal dimension and (b) a vertical dimension. The horizontal dimension pertains to the specialized networking requirements of the optical/MPLS environment. It indicates the enhancements necessary for the MPLS TE control plane model to address the peculiar optical networking requirements. The

vertical dimension pertains to localized hardware and software characteristics of the OXCs, which helps to determine the device-specific adap-
tations and support mechanisms needed to port and reuse the MPLS TE
control plane software artifacts on an OXC. Horizontal dimension considerations include the following aspects:

- What type of state information should be discovered and dissemi-
  nated to support path selection for optical channel trails? Such
  state information may include domain-specific characteristics of
  the optical network (encoded as metrics), such as attenuation, dis-
  persion (chromatic, PMD), etc. This aspect will determine the type
  of additional extensions that are required for IGP link state advertisements to distribute such information.
- What infrastructure will be used to propagate the control
  information?
- How are constrained paths computed for optical channel trails
  that fulfill a set of performance and policy requirements subject to
  a set of system constraints?
- What are the domain-specific requirements for setting up optical
  channel trails, and what are the enhancements needed to existing
  MPLS signaling protocols to address these requirements?

Vertical dimension requirements include the management of the
planes in the optical LSR. We covered this subject when discussing
Figure 10–3.

GENERALIZED MPLS USE IN OPTICAL NETWORKS

The method of distributing and binding labels between LSRs can vary.
The LDP can be used, and so can extensions to RSVP. GMPLS has been
developed to support MPLS operations in optical networks and can use
such optical technologies as time division (e.g., SONET ADMs), wavelengths, and spatial switching (e.g. incoming port or fiber to outgoing
port or fiber) [ASHW01]. This part of the chapter describes those parts of
GMPLS that pertain to optical networks.

MPLS assumes LSRs have a forwarding protocol that is capable of
processing and routing signals that have packet, frame, or cell boundaries. LSRs are assumed to be O/E/O devices. In contrast, GMPLS assumes LSRs are O/O/O devices that recognize neither packet nor cell

boundaries. Thus, the forwarding decision is based on time slots, wavelengths, or physical ports.

We must pause a moment here to clarify some terms used in GMPLS. GMPLS uses the following terms (and the remainder of this chapter uses the GMPLS terminology).

- Packet-switched capable (PXC): Processes traffic according to packet/cell/frame boundaries.
- Time-division multiplex capable (TDM): Processes traffic according to a TDM boundary, such as a SONET/SDH node.
- Lambda-switch capable (LSC): Processes traffic according to the optical wavelength.
- Fiber-switch capable (FSC): Processes traffic according to the physical interface, such as an optical fiber.

Traditional MPLS LSPs are unidirectional, but GMPLS supports the establishment of bidirectional LSPs. Bidirectional LSPs have the benefit of lower setup latency and the requirement for fewer messages to support a setup operation.

Considerations for Interworking Layer 1 Lambdas and Layer 2 Labels

Optical switching is considered a layer 1 switching function, and MPLS operates at layer 2. Layer 1 switching is typically circuit-switched based, and layer 2 switching is packet-switched based. These differences are important to understanding how to interwork layer 1 lambda switching and layer 2 label switching and how these differences affect the following cases.

- Circuit-switched nodes may have thousands of physical links (ports). A key issue for optical/MPLS networks is the configuration and management of these ports and the wavelengths on the ports. Insofar as possible, it is desirable not to execute O/E/O functions in the core network in the data plane. Therefore, the ideal lightpath through the network would have use of the same wavelength end-to-end. This is not a trivial task, since it requires all nodes in one or more networks to agree on the specific wavelength. Notwithstanding, GMPLS permits the negotiation of these wavelengths.

Note: At this writing, it is not clear if the optical switches are going to need switching matrices that support these thousands of ports. It appears the photonic switching technology is being pushed into the future, and very large photonic cross-connects have not reached a point of high demand.

- The layer 1 switch ports do not have IP addresses. The intent is to use IP addresses for all nodes and the nodes' interfaces. This requirement adds a significant task to the optical/MPLS transport network.

- Layer 1 neighbor nodes do not need to know about their neighbor's internal port number ID; they need to know the channel ID on the port in order to recognize each piece of traffic. This condition still holds when MPLS is added to the mix, and I recommend a per-interface label assignment (as opposed to a per-platform (per switch) arrangement) in order to more easily meld with current layer 1 practices.

- Many of the circuit switches' features are configured manually, and the operations remain static (fixed slots, etc.). This practice cannot continue if the network resources are to be dynamically and adaptively utilized. Therefore, the optical/MPLS transport network adapts a new concept: The network no longer consists of fixed pipes; it is now dynamically changing. A good analogy is offered by [XU01]: The transport network can be thought of as a large circuit switch with a dynamically configurable backplane. Thus, the layer 1 operations must be amenable to the same kinds of bandwidth and OAM manipulation as the upper layers (such as ATM and MPLS at layer 2).

- The switching technology on circuit switches is based on a very fast hardware-oriented cross-connect fabric, wherein the input and output ports are tightly synchronized. In an optical/MPLS node, the next hop should be set up by the binding of the MPLS label in the cross-connect fabric to the output port to that next node. Furthermore, label distribution, say, with LDP, is analogous to, say, SS7 (ISUP) setting up connections at layer 1.

Examples of GMPLS Operations

This part of the chapter explains the GMPLS messages used in an optical network. The message that conveys this information can be sent to an optical LSR by a variety of protocols, such as extensions to RSVP and LDP. Consequently, we need not delve into the specific formats (bit

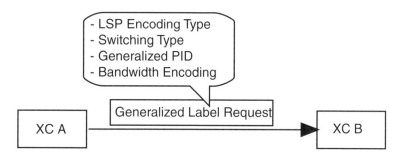

Figure 10–6   Generalized label request.

positions, and so on) here but can confine ourselves to understanding the functions of the fields in the message.

Generalized Label Request.   Figure 10–6 shows one example of how GMPLS is employed. XC A sends a GMPLS message to XC B. This message contains a control label called a Generalized Label Request. It provides an XC B with sufficient information to set up resources for the connection of an LSP with XC A. As depicted in Figure 10–6, three fields are required and one field is optional in this message .

These fields perform the following functions.

The first field is the LSP encoding type. It identifies the encoding (format) type to be used with the data associated with the LSP. This field tells the receiving node about the specific framing format. Thus far, 12 types have been stipulated, as shown in Table 10–1. Note that the

Table 10–1   LSP Encoding Type

Value	Type of Encoding (Format of Traffic)
1	Packet (conventional IP formats)
2	Ethernet V2
3	ANSI PDH (DS1, etc., payloads)
4	ETSI PSH (E1, etc., payloads)
5	SDH ITU-T G.707 (1996)
6	SONET ANSI T1.105 (1995)
7	Digital Wrapper
8	Lambda (photonic)
9	Fiber
10	Ethernet IEEE 802.3
11	SDH ITU-T G.707 (2000)
12	SONET ANSI T1.105 (2000)

Table 10–2   Generalized PID (G-PID)

Value	Type	Technology
0	Unknown	All
1	DS1 SF (Superframe)	ANSI-PDH
2	DS1 ESF (Extended Superframe)	ANSI-PDH
3	DS3 M23	ASNI-PDH
4	DE3 C-Bit Parity	ANSI-PDH
5	Asynchronous mapping of E4	SDH
6	Asynchronous mapping of DS3/T3	SDH
7	Asynchronous mapping of E3	SDH
8	Bit-synchronous mapping of E3	SDH
9	Byte-synchronous mapping of E3	SDH
10	Asynchronous mapping of DS2/T2	SDH
11	Bit-synchronous mapping of DS2/T2	SDH
12	Byte-synchronous mapping of DS2/T2	SDH
13	Asynchronous mapping of E1	SDH
14	Byte-synchronous mapping of E1	SDH
15	Byte-synchronous mapping of 31 * DS0	SDH
16	Asynchronous mapping of DS1/T1	SDH
17	Bit-synchronous mapping of DS1/T1	SDH
18	Byte-synchronous mapping of DS1/T1	SDH
19	Same as 12, but in a VC-12	SDH
20	Same as 13, but in a VC-12	SDH
21	Same as 14, but in a VC-12	SDH
22	DS1 SF Asynchronous	SONET
23	DS1 ESF Asynchronous	SONET
24	DS3 M23 Asynchronous	SONET
25	DS3 C-bit parity asynchronous	SONET
26	VT	SONET
27	STS	SONET
28	POS – no scrambling, 16-bit CRC	SONET
29	POS – no scrambling, 32-bit CRC	SONET
30	POS – scrambling, 16-bit CRC	SONET
31	POS – scrambling, 32-bit CRC	SONET
32	ATM mapping	SONET/SDH
33	Ethernet	Lambda, Fiber
34	Ethernet	Lambda, Fiber

Table 10–2   Generalized PID (G-PID) *(continued)*

Value	Type	Technology
35	SONET	Lambda, Fiber
36	Digital wrapper	Lambda, Fiber
37	Lambda	Fiber

lambda (photonic) encoding type identifies wavelength swathing (needing the services of an LSC module). The fiber encoding type identifies an FSC-capable module.

The second field is the switching type; it informs the XC about the type of switching that is to be performed on a particular link. The switching capabilities are defined as (a) several variations for PSC, (b) layer-2 switching capable (for example, ATM and Frame Relay), and (c) TDM switching capable, LSC switching capable, or FSC switching capable. The multiple options for PSC allow the network operator to stipulate more than one packet-switched operation.

The third field is the generalized protocol ID (G-PID); it identifies the payload that is carried by an LSP; that is, the traffic from the client for the specific LSP. This field uses the standard Ethertype values as well as the values shown in Table 10–2.

The fourth field is a bandwidth encoding value. It defines the bandwidth for the LSP. Table 10–3 shows the recommended values for this field.

Generalized Label.   The generalized label request can be extended to identify not only the labels for the packets but also (a) a single wavelength within a waveband or fiber, (b) a single fiber in a bundle of fibers, (c) a single waveband within a fiber, or (d) a set of time slots within a fiber or a waveband. This extension can also identify conventional ATM or Frame Relay labels.

Port and Wavelength Labels.   The FSC and PSC can use multiple channels/links that are controlled by a single control channel. If such is the case, the port/wavelength label identifies the port, fiber, or lambda that is used for this purpose.

Wavelength Label.   This label groups contiguous wavelengths and identifies them with a unique waveband ID. Its function is to provide a tool for the switch to cross-connect multiple wavelengths as one unit. This label contains three fields:

Table 10–3   Bandwidth Values

Signal Type	Bit Rate (Mbit/s)
DS0	0.064
DS1	1.544
E1	2.048
DS2	6.312
E2	8.448
Ethernet	10.00
E3	44.736
DS3	44.736
STS-1	51.84
Fast Ethernet	100.00
E4	139.264
OC-3/STM-1	155.52
OC-12/STM-4	622.08
Gigabit Ethernet	1000.00
OC-48/STM-1	2488.32
OC-192-STM-64	9953.28
10 G-Ethernet - LAN	1000.00
OC-768/STM-256	39812.12

- Waveband ID: The unique ID (selected by the sending node) of the wavelengths; to be used on all subsequent, related messages
- Start label: The lowest-value wavelength in the waveband
- End label: The highest-value wavelength in the waveband

Suggested Labels for Wavelengths

GMPLS can be used to configure the PXC hardware. One method, called *suggested labels,* is used by an upstream PXC to notify its neighbor downstream of a label that is to be used (suggested) for a wavelength or a set of wavelengths.

The GMPLS specification defines the use of a label set to limit the label choices made between adjacent GMPLS nodes. Label sets are useful when an optical node is restricted to a set of wavelengths; obviously, not all optical nodes have the same capabilities. The other helpful aspect of negotiating labels in regard to wavelengths is that some optical nodes are O/E/O capable and others are O/O/O capable; this aspect of the node will necessarily dictate wavelength capabilities.

BIDIRECTIONAL LSPS IN OPTICAL NETWORKS

GMPLS defines the use of bidirectional LSPs that have the same traffic engineering requirements in each direction.[2] To establish a bidirectional LSP when RSVP-TE or CR-LDP is used, two unidirectional paths between peer LSRs must be independently established. The principal disadvantage to this approach is the time it takes to set up this bidirectional relationship. Second, setting up two unidirectional LSPs requires more messages to be exchanged than with the setting up of one symmetric bidirectional LSP. Third, independent LSPs for a user's traffic profile can lead to different routes taken through the network for the two LSPs. This situation may not be a problem, but it does make for more complex resource allocation schemes.

Label Contention Resolution

Of course, since either optical node can initiate label allocations, it is possible for two peer nodes to suggest, at about the same time, labels that are in conflict with each other (e.g., the same label values for different fibers or the same labels for different wavelengths on a fiber). This is not a major issue. Contention resolution of potentially conflicting virtual circuit or label bindings is well studied, and GMPLS defines the procedures for resolving these label contention conflicts.

Link Protection

Another attractive feature of GMPLS is the provision for link protection information, including the kind of protection that is needed. The following types of link protection are defined in GMPLS:

- Enhanced: Signifies that a protection scheme that is more reliable than dedicated 1+1 should be used.

---

[2] My clients and I have long favored the simultaneous setting up of bidirectional virtual circuits in X.25, Frame Relay, ATM, and for MPLS: LSPs. It is good news to us that GMPLS adopts this approach for the reasons cited in the text of this chapter. Of course, one idea behind two unidirectional connections/LSPs for a user's traffic is the understanding that many users' traffic flows are asymmetric, with more traffic flow in one direction than the other. The idea is to go find the bandwidth, wherever it is in the network, to support this asymmetric flow. Fine, but a multigigabit fiber network makes the argument for independent unidirectional traffic paths even less tenable.

- Dedicated 1+1: Signifies that a dedicated link layer protection scheme should be used.
- Dedicated 1:1: Signifies that a dedicated link layer protection scheme, i.e., 1:1 protection, should be used to support the LSP.
- Shared: Signifies that a shared link layer protection scheme, such as 1:N protection, should be used to support the LSP.
- Unprotected: Signifies that the LSP should not use any link layer protection.
- Extra Traffic: Signifies that the LSP should use links that protect other higher priority traffic. Such LSPs can be preempted when the links carrying the higher priority traffic being protected fail.

SUMMARY

The use of MPLS in the WDM-based optical network is not a given, but the attractive aspects of MPLS (traffic engineering, QOS negotiation, graceful integration of IP, etc.) provide a useful ally to the optical planes. [AWDU01] has established the model for multiprotocol lambda switching. [BLAC02] describes the details of the MPLS and optical control planes interactions.

GMPLS is considered a prime candidate for control plane management. The "G" (for generalized) is not quite accurate. GMPLS is tailored to support the specific interworking of MPLS with optical networks.

11

VPNs with L2TP, BGP, OSPF, and MPLS

This chapter explains how virtual private networks (VPNs) operate using L2TP, BGP, OSPF, and MPLS. The main focus is on a VPN model that uses BGP route and label dissemination, and MPLS label stacking for forwarding traffic across the VPN.

The first three sections of this chapter are brief tutorials on VPNs, L2TP, and BGP (and OSPF is explained in Chapter 8). The knowledgeable reader can skip these sections.

OVERVIEW OF VPNs

The term Virtual Private Network (VPN) has been in the industry for many years. It was first used in X.25-based public packet networks in the mid-1970s to describe a data network that was available to the public, much like the public telephone network was available to voice customers. For the data VPN, the idea was to provide service to the customer such that the customer perceived that the service was coming from a private network, one tailored to the customer's needs, when in fact the network was used by many other customers.

These ideas about the VPN still hold today. A modern VPN is a network or set of networks that allow multiple user organizations and user sites to share the services of network infrastructure. The same addressing,

QOS, and security are made available to the VPNs' customers as would be available in a private network.

The major characteristics of VPNs are as follows:

- The VPN user is not aware of the VPN bearer services, such as the Internet or a private internet. Many VPNs are implemented with private leased lines and private networks, but the user does not know about this infrastructure.
- The VPN user is provided with extensive security services, including privacy and authentication operations.
- The VPN user is also provided with extensive addressing services. The user is allowed to use private or public addresses. It is the responsibility of the VPN to make sure all addresses are unambiguous, even with the use of overlapping private addresses from two different users.
- The employees or customers of an organization can dial in to the VPN backbone and obtain automatic configuration services from remote servers.
- The VPN customer is "tunneled" through the VPN domain with headers that encapsulate the user traffic to provide (a) negotiation of services with servers (using PPP and L2TP), (b) protection of the user traffic (IPSec), (c) forwarding of traffic, and (d) provision of QOS (MPLS).

ROLE OF TUNNELS IN VPNs (BY USE OF L2TP)

The Point-to-Point Protocol (PPP), the Layer 2 Tunneling Protocol (L2TP), remote RADIUS servers, and IPSec become important tools to support VPNs. Many VPN features are executed with (say) L2TP tunnels, and increasingly, MPLS LSPs are being used as tunnels as well. The idea is illustrated in Figure 11–1.

L2TP was introduced to allow the use of the PPP procedures between different networks and multiple communications links. With L2TP, PPP is extended as an encapsulation and negotiation protocol to allow the transport of PPP and user traffic between different networks and nodes.

One principal reason for the advent of L2TP is the need to dial in to a network access server (NAS) that may reside at a remote location. Although this NAS can be accessed through the dial-up link, it may be that

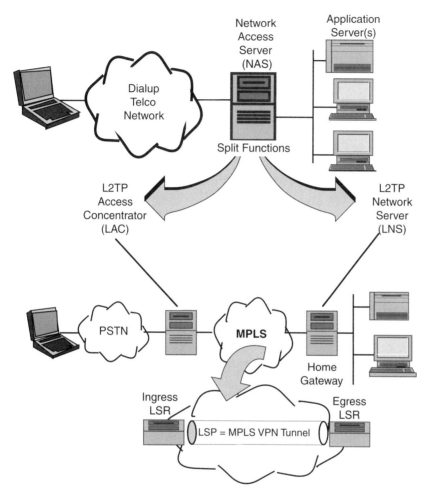

Figure 11–1   L2TP and MPLS interworking.

the NAS is located in another network. L2TP allows the use of all the
PPP operations that can be used between machines in different networks. With the implementation of L2TP, an end user establishes a
layer 2 connection to an access concentrator such as a modem bank, an
ADSL bank, etc. Thereafter, the concentrator is responsible for creating
a tunnel and sending the specific PPP packets to a network access server.

Before the advent of L2TP, these capabilities were proprietary. For
example, Microsoft developed the Point-to-Point Tunneling Protocol
(PPTP) and Cisco developed the Layer 2 Forwarding Protocol (L2FP).
L2TP is a standard that encompasses the attributes of these proprietary
protocols.

L2TP provides a number of other useful services. First, multiple protocols can be supported and negotiated, although IP is the prevalent protocol. L2TP also allows the use of unregistered IP addresses through the use of tunnels. A NAS can be used to assign addresses from a single address pool, thus simplifying the IP address management process. L2TP also permits the centralization of login and authentication operations by colocating a NAS with an L2TP Access Concentrator (LAC). L2TP allows a virtual dial-up service by which many autonomous system protocol domains share access to core components such as routers, modems, and access servers.

Using L2TP, the NAS becomes two machines, and its work is split between these machines. These machines communicate with an L2TP tunnel. These machines perform the following operations:

1. *L2TP Access Concentrator (LAC):* The LAC is the dial-in user side of the L2TP tunnel. It deals with the user's specific point-to-point connection. The LAC is responsible for tunneling and de-tunneling operations between the user and the LNS.

2. *L2TP Network Server (LNS):* The LNS is a node acting at one side of the peer L2TP Tunnel endpoint. Its other peer is the L2TP Access Concentrator (LAC). The LNS is the termination point of a PPP session that is being tunneled from the LAC. The NAS need not be concerned with the user-LAC operations. It need not support the physical and data link characteristics of the user computer; that is the job of the LAC.

3. *L2TP Tunnel:* This tunnel exists between the LAC and LNS peers. It consists of the user traffic and the header information necessary to support the tunnel. Therefore, the tunnel provides the encapsulated PPP datagrams and the requisite control messages needed for the operations between the LAC and LNS.

In Figure 11–1, we see the placement of the LAC and the LNS with respect to the public switch telephone network (PSTN) and an MPLS network. The basic concept of L2TP is for a remote system to initiate a PPP connection through the PSTN to a LAC. The LAC's job is to tunnel the PPP connection through the MPLS backbone network to a local LNS. At this LNS, a home local area network (LAN) is discovered. After the discovery process is completed, the traffic is delivered to the end user through the L2TP "tunnel."

MPLS comes into the picture in the VPN backbone. It supports the transport of L2TP tunnels (or other tunnels, such as IPSec tunnels) through the VPN nodes from the ingress LSR to the egress LSR. Thus, an MPLS LSP acts as a tunnel for, say, L2TP tunnels.

OVERVIEW OF BGP

The Border Gateway Protocol (BGP) is an inter-autonomous system protocol and is a relatively new addition to the family of routing protocols (it has seen use since 1989, but not extensively until the last few years). Today, it is the principal route advertising protocol used in the Internet for external gateway operations. Figure 11–2 shows a BGP topology and some key terms.

BGP is designed to run with a reliable transport layer protocol, such as TCP. Therefore, the BGP network manager need not be concerned about correct receipt of traffic, segmentation, etc. These potential problems are handled by the transport layer.

BGP operates by building a graph of ASs. The graph is derived from the routing information exchanged by the BGP routers in the ASs. BGP considers the entire Internet as a graph of ASs, with each identified by an AS number. The graph between the ASs is also called a tree. While autonomous systems are usually connected in a neighbor relationship, a BGP router can be configured to skip over intermediate routers in the AS path tree.

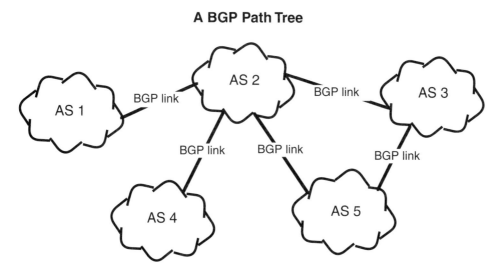

A BGP Path Tree

Figure 11–2   BGP AS links and path tree.

Here are some of the major features of BGP:

- BGP sends messages only if something changes, and not continuously. Obviously, this procedure keeps the overhead down on the link.
- BGP is able to select a loop-free path even though the system may contain physical loops.
- BGP stores backup paths, and in the event of failure of the primary path, it need not count to infinity waiting for the network routing tables to stabilize.
- Routing decisions can be based on policy considerations and need not be based just on the fewest number of hops. This point is important for public networks, like the Internet, in which the ISPs enter into peering arrangements with each other. These arrangements can be supported with BGP routing policies.
- A BGP router enters into a relationship with another BGP router through manual configurations, and not automatically. This is also important in the Internet to support or deny peering arrangements.

BGP Neighbors

Routing protocols need to know about their neighbors and how to exchange hellos and routing information with them. BGP is no exception, and this protocol also must consider factors beyond a hop count or a link metric. BGP must also deal with policy-based routing, since BGP may entail the transmission of traffic between different (and potentially competitive) enterprises. Therefore, BGP's neighbors are very important because they may be *external* neighbors, those belonging to another autonomous system and another enterprise.

From the technical standpoint, BGP supports two kinds of neighbors. The internal neighbor is in the same autonomous system, and the external neighbor is in a different autonomous system. As a general practice, external neighbors who are adjacent to each other a share a subnet. Internal neighbors can be nonadjacent physically; they can be located anywhere in the autonomous system.

BGP Speakers

BGP uses the concept of a *speaker* to advertise routing information. The speaker resides in the router. Using a common set of policies (routing agreements, described shortly), the BGP speakers arrive at an agreement as to which AS border routers will serve as exit/entry points for particular networks outside the AS. This information is communicated to the AS's internal routers, through the interior routing protocol or with manual configurations.

Connections between BGP speakers of different ASs are called *external* links. BGP connections between BGP speakers within the same AS are called *internal* links. A peer in a different AS is referred to as an external peer, and a peer in the same AS is described as an internal peer.

Communities

BGP can be set up to distribute routing information about a group of destinations (networks) called communities. The idea is to be able to group destinations into these communities and apply routing policies to a community. This approach simplifies the speaker's job by aggregating routing information. It also provides a tool for the network manager to control the dissemination of routing information.

There is considerable flexibility in using the BGP community operation. For example, a destination can belong to multiple communities, and the AS network manager can define the community to which a destination belongs. Given that one belongs or does not belong to a community, BGP will or will not support the distribution of routing information.

BGP POLICY-BASED ARCHITECTURE

As stated earlier, one of the unique aspects of BGP is its policy-based architecture. RFC 1655 describes this architecture, and this part of the chapter summarizes the RFC 1655 description of this aspect of BGP.

BGP provides the capability for enforcing policies based on various routing preferences and constraints, such as economic, security, or political considerations. Policies are not directly encoded in the protocol. Rather, policies are provided to BGP in the form of configuration information, explained in the last section of this chapter. The BGP routers allow the network manager (the AS administration) to perform "policy configuration" tasks when a policy is created or changed. These changes affect the path selection at the router, as well as the redistribution of routing in the BGP domain.

Of course, efficient traffic management is important, and BGP can control the following aspects of traffic forwarding at the router:

- An AS can minimize the number of transit ASs. (Shorter AS paths can be preferred over longer ones.)
- If an AS determines that two or more AS paths can be used to reach a given destination, the AS can use a variety of means to decide which of the candidate AS paths it will use. The quality of an AS can be measured by such things as link speed, capacity, and congestion tendencies. Information about these qualities might be determined by means other than BGP, such a router's administrative metrics.
- Preference of internal routes over external routes.

PATH SELECTION WITH BGP

A BGP speaker evaluates different paths to a destination network from its border gateways at that network, selects the best one, applies relevant policy constraints, and then advertises it to all of its BGP neighbors. A complication in inter-AS routing does arise from the lack of a universally agreed-upon metric among ASs that can be used to evaluate external paths. Each AS may have its own set of criteria for path evaluation.

Anyway, the BGP speaker builds a routing database consisting of the set of all feasible paths and the list of networks reachable through each path. In actual BGP implementations, the criteria for assigning degrees of preferences to a path is specified in configuration tasks. These tasks include configuring BGP neighbors, setting up administrative weights for a path to a neighbor, restricting routing information to and from neighbors, setting up aggregate addresses, etc.

MPLS LABEL STACKING IN THE VPN

Figure 11–3 shows how MPLS could be used to support a large base of VPN customers with a very simple arrangement. Certain assumptions must be made for this operation to work well. First, the customers are at the same ends of the MPLS tunnel. Second, they have the same QOS requirements. But these two requirements, should not be unusual. For example, many customers may be running VoIP from one site to another.

Three sets of customers are supported by the VPN in this example (A, B, and C), but there could be hundreds or even thousands. The point is that by label stacking, the VPN backbone can accommodate all the

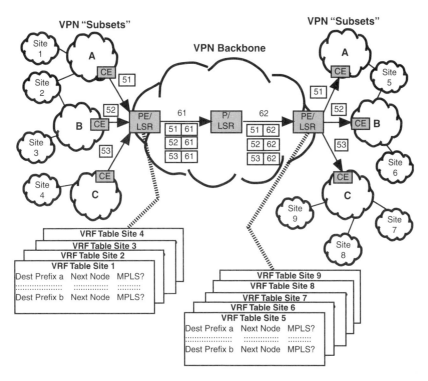

Figure 11–3   Label stacking in a VPN.

traffic with one set of labels for the LSP in the backbone. In Figure 11–3, the customers' labels (labels 51, 52, 53) are pushed down and not examined through the MPLS tunnel. The MPLS tunnels use labels 61 and 62.

VPN ARCHITECTURE

There are many ways to design a VPN. The remainder of this chapter (using Figure 11–3) discusses a VPN architecture published in [ROSE01b], [XU01], and [BERN01]. Many of the concepts expressed in [ROSE01b] are implemented in Cisco routers.

Customer Edge (CE) Devices

As shown in Figure 11–3, at each customer site, are one or more customer edge (CE) devices, each of which is "attached" by a data link (e.g., PPP, ATM, Ethernet, Frame Relay, etc.) to one or more LSR-based provider edge (PE/LSR) routers. Routers in the provider's network that

do not attach to CE devices are known as P/LSR routers. In general, the
CE device can be expected to be a router, which [ROSE01b] calls the CE
router.

A PE router is attached to a particular VPN if it is attached to a CE
device in that VPN. A PE router is attached to a particular site if it is attached to a CE device in that site. When the CE device is a router, it is a
routing peer of the PE(s) to which it is attached, but it is not a routing
peer of CE routers at other sites. Routers at different sites do not directly
exchange routing information with each other. As a consequence, the
customer has no backbone or "virtual backbone" to manage and does not
have to deal with any intersite routing issues.

Clear administrative boundaries are maintained between the service provider (SP) and its customers. Customers are not required to access the PE or P routers for management purposes, nor is the SP
required to access the CE devices for management purposes.

One rule for the [ROSE01b] architecture is that two CE sites can
communicate with each other over the VPN backbone only if the sites are
part of the VPN "set" of sites. The rule is quite simple; if a site is not part
of the VPN, it cannot avail itself of the VPN backbone. Therefore, this architecture is oriented toward private, corporate internets and the public
Internet.

Multiple Forwarding Tables

One of the key aspects of VPNs is the use of multiple forwarding tables, known as VPN routing and forwarding tables (VRFs). Each PE
router maintains these tables, as shown in Figure 11–3. Every site to
which the PE is attached must be mapped to one of the forwarding tables. When a packet is received from a particular site, the forwarding
table associated with that site is accessed to determine how to route the
packet. The forwarding table associated with a particular site S is populated only with routes that lead to other sites that have at least one VPN
in common with S. This prevents communication between sites that have
no VPN in common and is a very important aspect of VPN security.

A PE router is attached to a site by virtue of being the endpoint of
an interface whose other endpoint is a CE device. If there are multiple
attachments between a site and a PE router, all the attachments can be
mapped to the same forwarding table or different attachments can be
mapped to different forwarding tables. When a PE router receives a
packet from a CE device, it knows the interface over which the packet arrived, and this determines the forwarding table used for processing that

packet. The choice of a forwarding table is not determined by the user content of the packet. Different sites can be mapped to the same forwarding table, but only if they have all their VPNs in common.

The use of separate forwarding tables allows the reuse of the RFC 1918 private addresses. In addition (and explained shortly), a VPN route distinguisher ID is added to the IP address to ensure that there is no ambiguity of IP addresses and prefixes.

A PE router will also have a default forwarding table, which is not associated with any particular VPN site or sites. The default forwarding table handles traffic that is not VPN traffic, as well as VPN traffic that is simply transiting this router.

ROLE OF BGP IN VPNs

PE routers use BGP to distribute VPN routes to each other. The [ROSE01b] model allows each VPN to have its own address space, which means that a given address may denote different systems in different VPNs. If two routes to the same IP address prefix are actually routes to different systems, it is important to ensure that BGP not treat them as comparable. Otherwise BGP might choose to install only one of them, making the other system unreachable. Also, the value in BGP's POLICY field determines which packets get sent on which routes; given that several such routes are installed by BGP, only one such must appear in any particular VRF.

The VPN-IP4 Address Family

RFC 2858 defines an extension for BGP to carry routes for multiple address families, such as Systems Network Architecture (SNA), and OSI addresses. This extension is used in VPNs to define the VPN-IP4 address family. It is an 11-byte field with an 8-byte route distinguisher (RD), followed by the 4-byte IPv4 address. This new address ensures that if the same IPv4 address is used in two different VPNs, it will be possible to install two different routes to that address, one for each VPN.

Using BGP to Distribute the Address and the Label

If two sites of a VPN attach to PEs that are in the same autonomous system, the PEs can distribute VPN-IPv4 routes to each other by means of an IBGP connection between them. Alternatively, each can have an IBGP connection to a route reflector.

When a PE router distributes a VPN-IPv4 route via BGP, it uses its own address as the "BGP next hop." This address is encoded as a VPN-IPv4 address with an RD of 0. It also assigns and distributes an MPLS label. As shown in Figure 11–3, when the PE processes a received packet that has this label at the top of the stack, the PE will pop the stack and process the packet appropriately. An important point is that the use of BGP-distributed MPLS labels is only possible if there is a label switched path between the PE router that installs the BGP-distributed route and PE router that is the BGP next hop of that route. Notice that the "user labels" of 51, 52, and 53 are not examined in the VPN backbone.

To ensure interoperability among systems that implement this VPN architecture, all such systems must support LDP, which is explained in Chapter 5.

Using Route Reflectors

Rather than having a complete IBGP mesh among the PEs, it is advantageous to make use of BGP route reflectors to improve scalability. Route reflectors are the only systems that need to have routing information for VPNs to which they are not directly attached. There are a number of options as to how route reflectors are used in MPLS and VPNs; these options are beyond our overview, and for more details, I refer you to section 4.3.3 of [ROSE01b].

ROLE OF OSPF IN VPNs

[ROSE01d] defines the operations for OSPF support of VPNs. This working draft contains many detailed rules for OSPF interworking with VPNs. This part of the chapter highlights this specification. If you are going to be using OSPF in your VPN, [ROSE01a] is an indispensable guideline for you.

Disadvantages and Advantages of Using OSPF in the VPN

The disadvantage to using OSPF on the VPN PE-CE link is that it gets the PE router involved in a VPN site's IGP (OSPF). The advantages are that (a) the administrators of the CE router need not have any expertise in any routing protocol other than OSPF, (b) the CE routers do not need to support any routing protocols other than OSPF, and (c) if a customer is transitioning a network from a traditional OSPF backbone to the VPN service, the use of OSPF on the PE-CE link eases the transitional issues.

Principal Rules for Using OSPF in VPNs

If the PE uses OSPF to distribute routes to the CE router, the standard procedures governing BGP/OSPF interactions would cause routes from one site to be delivered to another as AS-external routes (in type 5 LSAs). A better approach is to deliver such routes in type 3 LSAs (as interarea routes) so that they can be distinguished from any "real" AS-external routes that may be circulating in the VPN. Hence, it is necessary for the PE routers to implement a modified version of the BGP/OSPF interaction procedures.

If two PEs attach to different VPN sites that are in the same OSPF area (as indicated by the OSPF area number), the PE/CE links to those site can be treated as links within that area. In addition, each PE may flood, into that area, a type 1 LSA advertising a link to the other PE. If this procedure is followed, two VPN sites in the same OSPF area will see the VPN backbone as a link within that area, a link between the two PE routers. [ROSE01d] calls this link a "sham link." This practice allows routes from one site to the other to be treated as intra-area routes.

Every PE attached to a particular OSPF network must be an OSPF area 0 router. This rule allows it to distribute interarea routes to the CE through type 3 LSAs. The CE router might or might not be an area 0 router, and the PE/CE link might or might not be an area 0 link.

If the OSPF network contains area 0 routers (other than the PE routers), at least one PE router must have an area 0 link to a non-PE area 0 router in that OSPF network. (The non-PE area 0 router functions as a CE router.) This ensures that interarea routes and AS-external routes can be leaked between the PE routers and the non-PE OSPF backbone.

When a type 3 LSA is sent over an area 0 link from a PE router to a CE router, a special bit in the LSA Options field is set. This ensures that if any CE router sends this type 3 LSA to a PE router, the PE router will not further redistribute it.

Two sites that are not in the same OSPF area will see the VPN backbone as being an integral part of the OSPF backbone. However, if there are area 0 routers that are NOT PE routers, then the VPN backbone actually functions as a higher-level backbone, providing a third level of hierarchy above area 0. This approach allows a legacy OSPF backbone to become disconnected during a transition period, as long as the various segments all attach to the VPN backbone.

If a PE and a CE are communicating through OSPF, the PE must create, and must flood to the CE, a type 1 LSA advertising its link to the CE. The PE must have an OSPF router ID that is valid (i.e., unique) within the OSPF domain. The PE must also be configured to know which OSPF area the link is in. A PE-CE link can be in any area, including area 0.

The PE must support at least one OSPF instance for each OSPF domain to which it attaches. Each instance of OSPF must be associated with a single VRF. If n CEs associated with that VRF are running OSPF on their respective PE/CE links, then those n CEs are OSPF adjacencies of the PE in the corresponding instance of OSPF.

If the OSPF domain has any area 0 routers (other than the PE routers), then at least one of those *must* be a CE router and must have an area 0 link to at least one PE router. This adjacency may be through an OSPF virtual link.

As stated, these are just the highlights of the OSPF/VPN rules. Again, see [ROSE01d] for the details.

ROLE OF MPLS IN VPNs

When a PE receives a packet from a CE device, it chooses a particular VRF in which to look up the packet's destination address. As noted earlier, this choice is based on the packet's incoming interface. Assume that a match is found. As a result we learn a "next hop" and an outgoing interface.

Keeping VPN Routers Out of P Routers

The packet must travel at least one hop through the backbone. The packet thus has a "BGP next hop," and the BGP next hop will have assigned a label for the route that best matches the packet's FEC. This label is pushed onto the packet's label stack and becomes the bottom label. In Figure 11–3, the label might be 51.

The packet will also have an "IGP next hop," created with OSPF, IS-IS, or RIP, which is the next hop along the IGP route to the BGP next hop. The IGP next hop will have assigned a label for the route that best matches the address of the BGP next hop. This label gets pushed on as the packet's top label. In Figure 11–3, this label is 61. The packet is then forwarded to the IGP next hop. The process continues until the end of the LSP tunnel is reached.

Once again, it is the two-level labeling that makes it possible to keep all the VPN routes out of the P routers, and this in turn is crucial to ensuring the scalability of the VPN. The backbone does not even need to have routes to the CEs, only to the PEs.

Isolating the VPNs

To maintain proper isolation of one VPN from another, no router in the backbone should accept a labeled packet from any adjacent non-backbone node unless the following two conditions hold:

1. The label at the top of the label stack was actually distributed by that backbone router to that non-backbone node.
2. The backbone router can determine that use of the label will cause the packet to leave the backbone before any labels lower in the stack will be inspected and before the IP header will be inspected.

The first condition ensures that any labeled packets received from non-backbone routers have a legitimate and properly assigned label at the top of the label stack. The second condition ensures that the backbone routers will never look below that top label.

SUMMARY

MPLS is a powerful tool for the VPN service provider. By the simple use of BGP extensions, route reflectors, and the "BGP next hop" feature, MPLS allows the VPN to scale well beyond what it could do without MPLS.

Other protocols can be helpful and should be used as part of the VPN tool box. These include PPP, L2TP, and OSPF.

12

MPLS and DiffServ

This chapter explains how MPLS networks support differentiated services (DiffServ or DS) operations. It describes how DiffServ behavior aggregates (BAs) are mapped onto label switched paths (LSPs) in order to best match DiffServ traffic engineering and QOS objectives.

DIFFSERV CONCEPTS

The main ideas of DiffServ are (a) to classify traffic at the boundaries of network and (b) to regulate (condition) this traffic at the boundaries. The classification operation entails the assignment of the traffic to behavioral aggregates (BAs). These behavioral aggregates are a collection of packets with common characteristics in how they are identified and treated by the network. The network classifies the packets according to the content of the packet headers. The idea is to have a small number of classifications to simplify the allocation of resources for the traffic classes.

The identified traffic is assigned a value, a differentiated services codepoint. For IPv4, the codepoint is the use of 6 bits of the TOS field; for IPv6, the codepoint is in the traffic class octet. Since DiffServ assumes the use of just a few traffic classes, the "redefined" TOS field is considered sufficient to handle the different classes (64 of them).

Per-Hop Behavior

After the packets have been classified at the boundary of the network, they are forwarded through the network in accordance with the DS codepoint (DSCP). The forwarding is performed on a per-hop basis; that is, the DS node alone decides how the forwarding is to be carried out. This concept is called *per-hop behavior* (PHB). At each node, the DSCP is used to select the PHB, which in turn determines the scheduling treatment for the packet and possibly the drop probability for the packet.

THE DIFFSERV DOMAIN

DiffServ uses the idea of a DS domain, shown in Figure 12–1. A collection of networks operating under an administration could be a DS domain, say, an ISP. It is responsible for meeting a service level agreement (SLA) between the user and the DS domain service provider.

The DS domain consists of a contiguous set of nodes that are DS compliant and agree to a common set of service provisioning policies. The DS domain also operates with a common per-hop behavior definition (more than one PHB is allowed; the aggregate is called a PHB group). The PHB defines how a collection of packets with the same DS codepoint is treated.

The DS domain contains DS boundary nodes that are responsible for the classifying operations and the conditioning of ingress traffic. I will

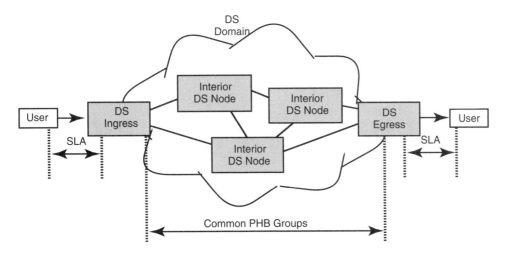

Figure 12–1   DiffServ domain.

have more to say about conditioning later; for this introduction, it consists of controlling the traffic to make sure it "behaves" according to the rules of the DS domain (and, one hopes, the desires of the user).

Once past the ingress node and inside the DS domain, the internal nodes forward packets based on the DS codepoint. Their job is to map the DS codepoint value to a supported PHB. Thus, there are DS boundary nodes and DS interior nodes. The DS boundary nodes connect the DS domain to other DS domains or noncompliant systems. There is no restriction on what type of machine executes the boundary or interior node operations. For instance, a host might play the role of a DS boundary node.

TYPES OF PER-HOP BEHAVIORS

Three types of PHBs are defined in the DS specifications. DiffServ defines a default PHB in which no special treatment is accorded to the packet. It also defines expedited forwarding (EF), a method in which certain packets are given low-delay and low-loss service. Typically, these packets are regulated such that their queues are serviced at a rate in which the packets are removed from the buffer at least as quickly as packets are placed into the buffer.

The third PHB definition is assured forwarding (AF) [HEIN99]. This PHB is a tool that offers different levels of forwarding assurances for IP packets received from a user. (The WFQ operations discussed in Chapter 7 would be good tools for managing AF traffic).

Four AF classes are defined, where each AF class in each DS node is allocated a certain amount of forwarding resources (buffer space and bandwidth). Within each AF, class packets are marked (again by the user or the service provider) with one of three possible drop-precedence values. Thus, the number of AF PHBs is 12.

In case of congestion, the drop precedence of a packet determines the relative importance of the packet within the AF class. A congested DS node tries to protect packets with a lower drop-precedence value from being lost by preferably discarding packets with a higher drop-precedence value.

Within each AF class, an IP packet is assigned one of M different levels of drop precedence. A packet that belongs to an AF class i and has drop precedence j is marked with the AF codepoint AFij, where $1 \leq i \leq N$ and $1 \leq j \leq M$. Currently, four classes (N = 4) with three levels of drop precedence in each class (M = 3) are defined for general use.

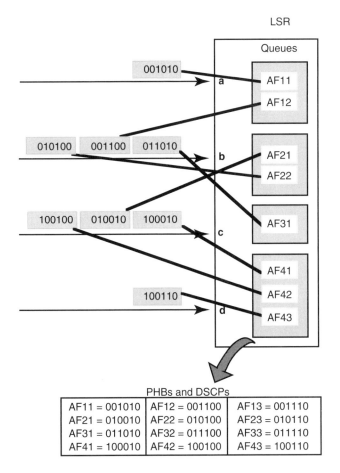

Figure 12–2   DCSPs and PHBs.

Figure 12–2 shows how AF queues are set up at a DS LSR. The value i identifies the queue for the packet, with each queue holding a class of traffic. The value j determines the drop preference of the packets belonging to the same queue. Four queues are set up at the DS LSR, and four incoming links deliver packets across interfaces a, b, c, and d. The DSCPs in the packets are used to place the packets in their respective queues, and the AF PHB operations at this node determine how the packets are handled.

Note that the packets within each of the four queues are distinguished only by their drop preference. Of course, each queue is handled differently. Also note that the table at the bottom of the figure is used at the LSR to correlate the DSCP to a PHB, thus determining how the packet is treated at the LSR.

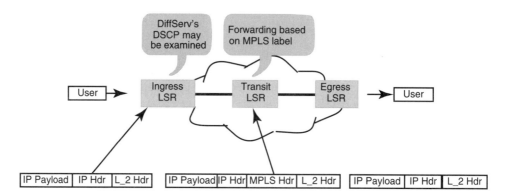

Figure 12–3   MPLS and DiffServ.

MPLS AND DIFFSERV ROUTERS

Considerable work is going on in the Internet working groups to define the relationships between MPLS and DiffServ, and the emphasis in this chapter is on [FAUC00]. Recall that DiffServ redefines the IPv4 TOS field and names it the DS codepoint. This field does not have to be processed by the MPLS transit routers, but it must be visible to the ingress and egress LSRs.

As shown in Figure 12–3, the egress router can use the DSCP to make decisions about how to code the MPLS label. Therefore, the label selection can determine how the traffic is treated in the network if the DSCP is used to determine the label.

The format for the MPLS header is shown (again) in Figure 12–4. It consists of the following fields:

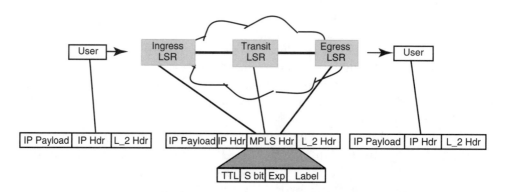

Figure 12–4   Format for the MPLS header.

- *Label*: Label value, 20 bits. This value contains the MPLS label.
- *EXP*: Experimental use, 3 bits. This field is not yet fully defined. Several Internet working papers on DiffServ discuss its use with this specification.
- *S*: Stacking bit, 1 bit. Used to stack multiple labels; discussed earlier in this book.
- *TTL*: Time to live, 8 bits. Places a limits on how many hops the MPLS PDU can traverse. This limit is needed because the IP TTL field is not examined by the transit LSRs.

TRAFFIC CLASSIFICATION AND CONDITIONING

The DS node must provide traffic classification and conditioning operations, as shown in Figure 12–5. The job of packet classification is to identify subsets of traffic that are to receive differentiated services by the DS domain. Classifiers operate in two modes: (a) the behavior aggregate classifier classifies packets only on the DS codepoint; (b) the multifield classifier classifies packets by multiple fields in the packet, such as codepoints, addresses, and port numbers. The BA defines all traffic crossing a link and requiring the same DS behavior.

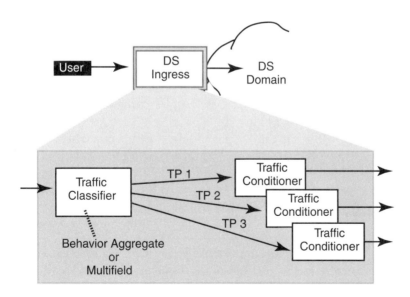

Figure 12–5   DiffServ classifiers and conditioners.

DS Classifiers

The classifiers provide the mechanism to guide the packets to a traffic conditioner for more processing. The traffic stream selected by the classifier is based on a specific set of traffic profiles, such as variable or constant bit rates, jitter, and delay.

The packets that are presented to a specific traffic conditioner constitute a traffic profile (TP) and can be in-profile or out-of-profile. In-profile means the packets are "conformant" to the user-network SLA. Out-of-profile packets are outside an SLA or, because of network behavior, arrive at the traffic conditioner at a rate that requires the conditioner to condition them (delay their delivery, drop them, etc.).

As a general practice, classification and conditioning operations take place at the network boundaries. Nothing precludes the internal nodes from invoking these operations, but their classification and conditioning operations are probably more limited than those of the boundary nodes.

Behavior Aggregates, Ordered Aggregates, and LSPs

The job of the MPLS network is to select how the DS BAs are mapped in the MPLS LSPs. As the MPLS network does this, the network administrator must be aware of another DS concept: the ordered aggregate (OA). The OA is a set of BAs that share an ordering constraint.

This term encompasses a DS rule that states that packets belonging to the same flow cannot be misordered if they differ only in drop precedence; that is, they must maintain the same order from the ingress LSR to the egress LSR. The effect of this rule is that packets belonging to the same set, such as AF21, AF22, and so on, are placed in a common queue for FIFO operations.

This idea is quite similar to ATM and Frame Relay. They, too, require that cells or frames belonging to a single virtual circuit cannot be misordered as they transit an ATM or Frame Relay network.

DS also defines the set of one or more PHBs that are applied to this set. The result of this definition is called a PHB scheduling class (PSC). The network administrator must decide if the sets of BAs are mapped onto the same LSP or different LSPs in one of two ways:

- With LSPs that can transport multiple ordered aggregates, so that the EXP field of the MPLS shim header conveys to the LSR the PHB to be applied to the packet (covering both information about the packet's scheduling treatment and its drop precedence).

- With LSPs that transport only a single ordered aggregate, so that the packet's scheduling treatment is inferred by the LSR exclusively from the packet's label value, while the packet's drop precedence is conveyed in the EXP field of the MPLS shim header or in the encapsulating link-layer-specific selective drop mechanism (ATM, Frame Relay).

Figure 12–6 shows a logical view of the relationships of the key DS functions for DS packet classification and traffic conditioning operations. The packets that exit the DS node in this figure must have the DS codepoint set to an appropriate value, based on the classification and traffic conditioning operations.

A traffic stream is selected by a classifier and sent to a traffic conditioner. DiffServ uses the term traffic conditioning block (TCB) to describe the overall conditioning operations. If appropriate, a meter is used to measure the traffic against a traffic profile. The results of the metering procedure may be used to mark, shape, or drop the traffic, based on the packet being in-profile or out-of-profile. The classifiers and meters can operate as a "team" to determine how the packet is treated with regard to marking, shaping, and dropping.

The packet marking procedure sets the DS field of a packet to a codepoint and adds the marked packet to a specific DS behavior aggregate. The marker can be configured to mark all packets steered to it to a single codepoint. Alternatively, the marker can mark a packet to one of a set of codepoints. The idea of this configuration is to select a PHB in a PHB group, according to the state of a meter. The changing of the codepoint is called *packet remarking*.

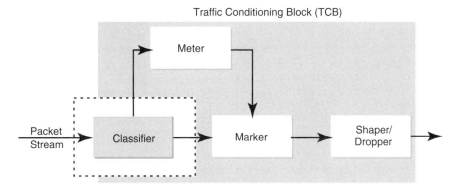

Figure 12–6   DS traffic classification and conditioning model.

The shaping procedure brings the packet stream into compliance with a particular traffic profile. The packet stream is stored in the shaper's buffer, and a packet can be discarded if there is not enough buffer space to hold a delayed packet.

The dropping procedure polices the packet stream to bring it into conformance with a particular traffic profile. It can drop packets to adhere to the profile. The figure shows the shaper and dropper as one entity because a dropper can be implemented as a special case of a shaper.

The originating node of the packet stream (the DS source domain) is allowed to perform classification and conditioning operations. This idea is called premarking and can be effective in supporting the end application's view of the required QOS for the packet stream. The source node may mark the codepoint to indicate high-priority traffic. Next, a first-hop router may mark this traffic with another codepoint, and condition the packet stream.

I stated that the collective operations of metering, marking, shaping, and dropping are known as the traffic conditioning block. The classifier need not be a part of the TCB, because it does not condition traffic. However, the classifiers and traffic conditioners can certainly be combined into the TCB. These options are shown in Figure 12–6 with the dashed lines.

The next part of this chapter goes into more detail on the classifiers and traffic conditioners. This material is available from the Internet draft authored by [BERN99].

CLASSIFICATION OPERATIONS

The main jobs of the classifier are to accept a packet stream (unclassified traffic) as input and to generate separate output streams (classified traffic). This output is fed into the metering or marking functions.

As mentioned earlier, the classifier operates as a behavior aggregate classifier or as a multifield (MF) classifier. The BA classifier uses only the DS codepoint to sort to an output stream, whereas the MF classifier uses other fields in the packet stream, such as a port number or an IP Protocol ID.

The BA or MF classifier checks are performed by filters, which are a set of conditions that are matched to the relevant fields in the packet to determine the output stream onto which the packet is placed. This idea is illustrated in Figure 12–7. Unclassified traffic is flowing into an interface

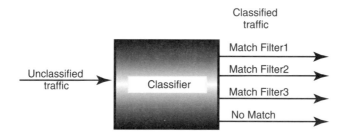

Figure 12–7   Example of the classifier's filtering operation.

and passed to the classifier. The filtering operations output the packets into four streams for the traffic-conditioning operations. The first three filters are exact matches on the BA or MA values. The no-match is a default filter to handle any packet types that have not been provisioned at the QOS node.

METERING OPERATIONS

After the packets have been classified, the meter monitors their arrival time in the packet stream to determine the level of conformance to a traffic profile. The profile has been preconfigured, perhaps according to an SLA contract.

Figure 12–8 is a functional diagram of the metering operation. The unmetered traffic is input to the metering function. This function is implemented with one or several types of meters defined in [BERN99], but others can be used: (a) the average rate meter, (b) the exponential weighted moving average meter, or (c) the token bucket meter. These meters are explained in Chapter 7.

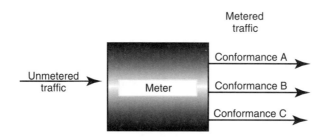

Figure 12–8   Example of the metering operations.

THE DS CODEPOINT REVISITED

As explained earlier, the DS field in the IP datagram is the IPv4 TOS field. It is called the DS codepoint (DSCP). This IPv4 8-bit field is shown in Figure 12–9, along with the redefinition according to the DS specifications.

The IPv4 *type of service (TOS)* field, shown in Figure 12–9(a), can be used to identify several QOS functions provided for an Internet application. Transit delay, throughput, precedence, and reliability can be requested with this field.

The TOS field contains five entries, consisting of 8 bits. Bits 0, 1, and 2 contain a precedence value that indicates the relative importance of the datagram. The next three bits are used for other services and are described as follows. Bit 3 is the *delay bit (D bit)*. When set to 1, this TOS requests a short delay through an internet. The aspect of delay is not defined in the standard, and it is up to the vendor to implement the service. The next bit is the *throughput bit (T bit)*. It is set to 1 to request high throughput through an internet. Again, its specific implementation is not defined in the standard. The next bit used is the *reliability bit (R bit)*, which allows a user to request high reliability for the datagram. The last bit of interest is the *cost bit (C bit)*, which is set to request the use of a

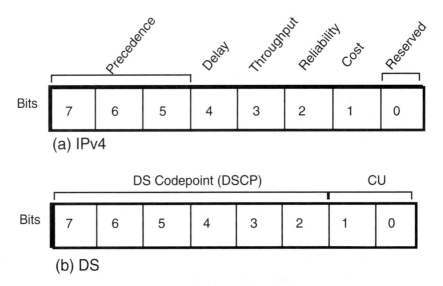

Figure 12–9   IPv4 TOS field and the DS codepoint.

low-cost link (from the standpoint of monetary cost). The last bit is not used at this time.

The DSCP is six bits in length, as depicted in Figure 12–9(b). The remaining two bits of the TOS field are currently unused (CU). The DSCP notation is xxxxxx, where x can be 1 or 0. The leftmost bit signifies bit 0 of the field, and the rightmost bit signifies bit 5. The entire 6-bit field is used by DS nodes as an index into a table to select a specific packet-handling mechanism.

The codepoints are related to the PHBs, and the PHBs include a default codepoint. A default configuration contains a recommended codepoint-to-PHB mapping. The default PHB is the conventional best-effort forwarding operation that exists today and is standardized in RFC 1812. When a link is not needed to satisfy another PHB, the traffic associated with the default PHB should be placed onto the link. RFC 2474 states that a default PHB should not be subject to bandwidth starvation and should be given some bandwidth, but the manner in which the bandwidth is provided is implementation specific. The available bit rate (ABR) operation in ATM is a good model to use for this implementation. The default codepoint for the default PHB is 000000.

The recommended codepoints can be amended or replaced with different codepoints at the discretion of the service provider. Even if the same PHBs are implemented on both sides of a DS boundary, the DSCP still can be re-marked.

If a DS node receives a packet containing an unrecognized codepoint, it simply treats the packet as if it were marked with the default codepoint. This rule means the DS node can examine other fields in the IP header (or layer 4 header) to know about the default codepoint. I make this point because this rule implies that a DS node must be able to review fields other than the codepoint. Strictly speaking, the fewer fields examined the better. If just the codepoint is examined, the operation can be very efficient, similar to ATM label switching.

The DSCP field can convey 64 distinct codepoints, as depicted in Table 12–1 and Figure 12–2. The codepoint space is divided into three pools for the purpose of codepoint assignment and management.

A pool of 32 codepoints (Pool 1) is assigned by Standards Action, as defined in the ongoing Internet standards. See Table 12–2. A pool of 16 codepoints (Pool 2) is reserved for experimental or local use (EXP/LU), and a pool of 16 codepoints (Pool 3) is initially available for experimental or local use. The DS standards state that Pool 3 should be preferentially utilized for standardized assignments if Pool 1 is exhausted.

Table 12–1   Recommended AF Code Points and AF Classes

	Class 1	Class 2	Class 3	Class 4
Low Drop Precedence	001010	010010	011010	100010
Medium Drop Precedence	001100	010100	011100	100100
High Drop Precedence	001110	010110	011110	100110

Table 12–2   Codepoint Assignments

Pool	Codepoint Space	Assignment Policy
1	xxxxx0	Standards Action
2	xxxx11	EXP/LU
3	xxxx01	EXP/LU (*)

(*) May be utilized for future allocations as necessary.

Codepoints for Assured Forwarding

The recommended codepoints for the four general-use assured forwarding (AF) classes are described in the next section and in Figure 12–2. These codepoints do not overlap with any other general-use PHB groups.

DSCPS AND LSR USE OF MPLS LABELS

Recall that an interior LSR does not examine the IP header. The ingress LSR can examine the IP header for information to create an LSP and associated labels. Some means must be available to correlate the DS PHB with the packet. At first glance, it seems an easy task. Simply map the information in the DSCP into the label or the EXP fields of the shim header. It is not quite so straightforward, and this part of the chapter explains why.

The DSCP in the TOS field is 6 bits in length, and the EXP field in the MPLS shim header is 3 bits in length. Early in the development of DiffServ, it was intended that the 3-bit EXP field would support DS operations. The obvious problem is that the numbers of bits in these fields differ. This situation occurred because the EXP field was defined before the DiffServ working group was set up and it was believed that 3 bits would suffice. In addition, the old precedence field (also 3 bits in length) had been the field that routers used for making QOS decisions.[1]

---

[1]The TOS field is still in use for QOS functions. For example, the Cisco router defines several procedures for prioritization of traffic by use of this field, a topic explained in Chapter 7.

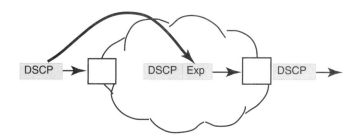

Figure 12–10   Mapping DSCP and EXP values to PHBs.

In addition, some designers think eight classes of traffic are sufficient to provide enough granularity for effective QOS operations. For that matter, a number of systems support only two classes of traffic, similar to the EF PHBs.

Moreover, as with any field, the working group wanted to keep the shim header small, and eight classes of traffic seemed appropriate. But, as noted in the previous discussion, the 6-bit DSCP allows up to 64 DSCPs.

Certainly, if a network supports eight or fewer PHBs, there is no problem. Let's assume a router is configured to support both MPLS and DiffServ operations. The following holds: (a) the DS-LSR maps DSCP values to suitable PHBs, and (b) the DS-LSR maps the EXP field values to suitable PHBs, a process depicted in Figure 12–10.

THE ORDERED AGGREGATE AND MPLS LSPS

Recall that an OA is a set of BAs that share an ordering constraint, and DS can define the set of one or more PHBs that are applied to this set. This ordering constraint has an interesting effect on MPLS LSPs. Since AF packets of the same class (say, AF21, AF22, etc.) must not be misordered, they should be assigned to the same LSP.

The result of this situation is the PHB scheduling class: a group of packets belonging to the same PHB must be sent over the same LSP. Two aspects of the PSC are of interest here: (a) the EXP-inferred-PSC LSPs and (b) the label-only-inferred-PSC LSPs. Let's take a look at these operations.

EXP-Inferred-PSC LSPs

By use of the 3-bit EXP field, a single LSP can support up to eight BAs of a given FEC. These LSPs are called EXP-inferred-PSC LSPs

(E-LSP), since the PSC of a packet transported on this LSP depends on the EXP field value for the packet. With this approach, the label can be used by the router to make forwarding decisions, and the EXP filed can be used to determine how to treat the packet.

Label-Only-Inferred-PSC LSPs

A separate LSP can be established for a single <FEC, OA> pair. With such LSPs, the PSC is explicitly signaled at label establishment time so that after label establishment, the LSR can infer exclusively from the label value the PSC to be applied to a labeled packet. This approach, inferring forwarding and ordering from the label, is a common implementation in ATM and Frame Relay networks.

When the shim header is used, the drop precedence to be applied by the LSR to the labeled packet is conveyed inside the labeled packet MPLS shim header with the EXP field. When the shim header is not used (such as MPLS over ATM or Frame Relay), the drop precedence is conveyed inside the link layer header encapsulation with link-layer-specific, drop-precedence fields (e.g., the ATM cell loss priority (CLP) bit or the Frame Relay discard eligibility (DE) bit).

This approach is called label-only-inferred-PSC LSPs (L-LSP) because the PSC can be inferred from the label without any other information (e.g., regardless of the EXP value).

Bandwidth Reservations for E-LSPs and L-LSPs

The Working Group defining MPLS support of DiffServ has developed the following guidelines for bandwidth reservations for E-LSPs and L-LSPs [FAUC00].

E-LSPs and L-LSPs can be established with or without bandwidth reservation. Establishing an E-LSP or L-LSP with bandwidth reservation means that bandwidth requirements for the LSP are signaled at LSP establishment time. Such signaled bandwidth requirements may be used by LSRs at establishment time to perform admission control of the signaled LSP over the DiffServ resources provisioned for the relevant PSC(s). Such signaled bandwidth requirements can also be used by LSRs at establishment time to perform adjustment to the DiffServ resources associated with the relevant PSC(s), for example, to adjust PSC scheduling weight.

When bandwidth requirements are signaled during the establishment of an L-LSP, the signaled bandwidth is obviously associated with the L-LSP's PSC. Thus, LSRs that use the signaled bandwidth to perform admission control can perform admission control over DiffServ resources that are dedicated to the PSC, for example, over the bandwidth guaranteed to the PSC through its scheduling weight.

When bandwidth requirements are signaled at establishment of an E-LSP, the signaled bandwidth is associated collectively with the whole LSP and therefore with the set of transported PSCs. Thus, LSRs that use the signaled bandwidth to perform admission control can perform admission control over global resources that are shared by the set of PSCs (e.g., over the total bandwidth of the link).

SUMMARY

In this chapter, we learned how MPLS networks support differentiated services (DiffServ or DS) operations. We also examined how DiffServ behavior aggregates are mapped onto label switched paths to best match DiffServ traffic engineering and QOS objectives.

Appendix A

Names, Addresses, Subnetting, Address Masks, and Prefixes

A newcomer to data networks is often perplexed when the subject of naming and addressing arises. Addresses in data networks are similar to postal addresses and telephone numbering schemes. Indeed, many of the networks that exist today have derived some of their addressing structures from the concepts of the telephone numbering plan.

It should prove useful to clarify the meaning of names, addresses, and routes. Table A–1 summarizes these definitions. A *name* is an identification of an entity (independent of its physical location), such as a person, an application program, or even a computer. An *address* is also an identification, but it reveals additional information about the entity, principally information about its physical or logical placement in a network. A *route* is information on how to relay traffic to a physical location (address).

A network usually provides a service that allows a network user to furnish the network with a name of something (another user, an application, etc.) that is to receive traffic. A network *name server* then uses this name to determine the address of the receiving entity. This address is then used by a routing protocol to determine the physical route to the receiver.

With this approach, a network user does not become involved with and is not aware of physical addresses and the physical location of other users and network resources. This practice allows the network administrator to relocate and reconfigure network resources without affecting

Table A–1   Names, Addresses, and Routes

Name An id of an entity, independent of physical location

    Example: JBrown@acme.com

Address An id that reveals a location of an entity

    Example: Network = 12.3, Subnetwork = 456, Host = 14

Route How to reach the entity at the address

    Example: Next node is Subnet 456

Practice is Name and address are correlated:

    12.3.456.14 is acme.com

end users. Likewise, users can move to other physical locations, but their names remain the same. The network changes its naming and routing tables to reflect the relocation.

PRINCIPAL ADDRESSES USED IN INTERNET AND INTRANETS

Communications between users through a data network require several forms of addressing. Typically, two addresses are required: (a) a physical address, also called a data link address, or a media access control (MAC) address on a LAN; and (b) a network address. Other identifiers, such as upper-layer names or port addresses, are needed for unambiguous end-to-end communications between two users.

Each device (such as a computer or workstation) on a communications link or network is identified with a physical address. This address is also called the hardware address. Many manufacturers place the physical address on a logic board within the device or in an interface unit connected directly to the device. Two physical addresses are employed in a communications dialogue; one address identifies the sender (source), and the other address identifies the receiver (destination). The length of the physical address varies, and most implementations use two 48-bit addresses.

The address detection operation on a LAN is illustrated in Figure A–1. Device A transmits a frame onto the channel. It is broadcast to all other stations attached to the channel, namely, stations B, C, and D. We assume that the destination physical address (DPA) contains the value C. Consequently, stations B and D ignore the frame. Station C accepts it, performs several tasks associated with the physical layer, strips

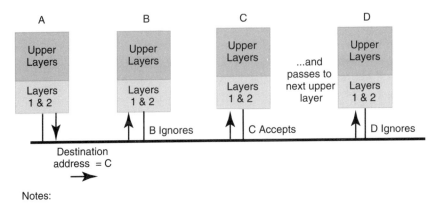

Figure A–1   Link address detection on a LAN.

away the physical layer headers and trailers, and passes the remainder of the protocol data unit (PDU) (it is no longer called a frame) to the next upper layer.

MAC Address

The IEEE assigns LAN addresses and universal protocol identifiers. Previously, Xerox Corporation did this work by administering what were known as block identifiers (block IDs) for Ethernet addresses. The Xerox Ethernet Administration Office assigned these values, which were three octets (24 bits) in length. The organization that received this address was free to use the remaining 24 bits of the Ethernet address in any way it chose.

Because of the progress made in the IEEE 802 project, it was decided that the IEEE would assume the task of assigning these universal identifiers for all LANs, not just CSMA/CD types of networks. However, the IEEE continues to honor the assignments made by the Ethernet Administration Office, although it now calls the block ID an *organization unique identifier (OUI)*.

The format for the OUI is shown in Figure A–2. The least significant bit of the address space corresponds to the individual/group (I/G) address

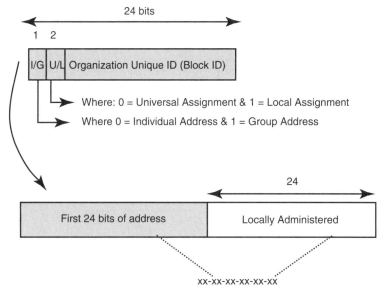

Figure A–2   The MAC address.

bit. The I/G address bit, if set to 0, means that the address field identifies an individual address. If the value is set to 1, the address field identifies a group address, which is used to identify more than one station connected to the LAN. If the entire OUI is set to all 1's, it signifies a broadcast address, which identifies all stations on the network.

The second bit of the address space is the local or universal bit (U/L). When this bit is set to 0, it has universal assignment significance—for example, it is from the IEEE. If it is set to 1, it is an address that is locally assigned. Bit position number two must always be set to 0 if the address is administered by the IEEE.

The OUI is extended to include a 48-bit universal LAN address (which is designated as the media access control address). The 24 bits of the address space is the same as the OUI assigned by the IEEE. The one exception is that the I/G bit may be set to 1 or 0 to identify group or individual addresses. The second part of the address space consisting of the remaining 24 bits is locally administered and can be set to any values an organization chooses.

Network Address

A network address identifies a network or networks. Part of the network address can also designate a computer, a terminal, or anything that a private network administrator wants to identify within a network (or attached to a network), although the Internet standards place very strict rules on what an IP address identifies.

A network address is a higher-level address than the physical address. The components in an internet that deal with network addresses need not be concerned with physical addresses until the data has arrived at the network link to which the physical device is attached.

This important concept is illustrated in Figure A–3. Assume that a user (host computer) in Los Angeles transmits packets to a packet network for relaying to a workstation on a LAN in London. The network in London has a network address of XYZ (this address scheme is explained shortly).

The packets are passed through the packet network (using the network's internal routing mechanisms) to the packet switch in New York. The packet switch in New York routes the packet to the gateway located in London. This gateway examines the destination network address in the packet and determines that the packet is to be routed to network XYZ. It then transmits the packet onto the appropriate communications channel (link) to the node on the LAN that is responsible for communicating with the London gateway.

Notice that this operation did not use any physical addresses in these routing operations. The packet switches and gateway were only concerned with the destination network address of XYZ.

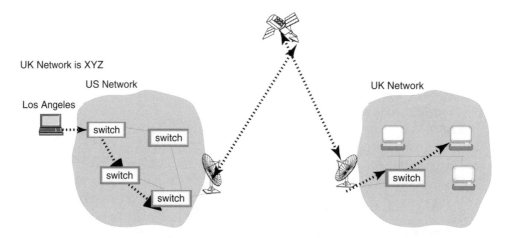

Figure A–3   Network layer addressing.

You might question how the London LAN is able to pass the packet to the correct device (host). As we learned earlier, a physical address is needed to prevent every packet from being processed by the upper-layer network layer protocols residing in every host attached to the network. Therefore, the answer is that the target network (or gateway) must be able to translate a higher-layer network destination address to a lower-layer physical destination address.

Figure A–4, a node on the LAN, is a server that is tasked with address resolution. Let us assume that the destination address contains a network address, such as 128.1, *and* a host address, say, 3.2. Therefore, the two addresses could be joined (concatenated) to create a full internet network address, which would appear as 128.1.3.2 in the destination address field of the IP datagram.

Once the LAN node receives the datagram from the gateway, it must examine the host address and either (a) perform a lookup into a table that contains the local physical address and its associated network address or (b) query the station for its physical address. Then, it encapsulates the user data into the LAN frame, places the appropriate LAN physical layer address in the destination address of the frame, and transmits the frame onto the LAN channel. All devices on the network examine the physical address. If this address matches the device's address, the PDU is passed to the next upper layer; otherwise, it is ignored.

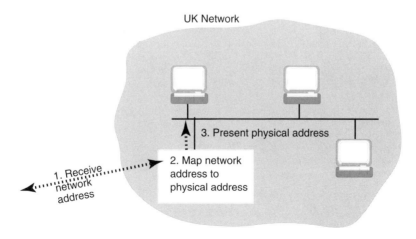

UK Network

3. Present physical address

2. Map network address to physical address

1. Receive network address

Figure A–4 Mapping network addresses to physical addresses.

IP Address

IP networks use a 32-bit, layer 3 address to identify a host computer and the network to which the host is attached. The structure of the IP address is depicted in Figure A–5. Its format is:

IP Address = Network Address + Host Address

The IP address identifies a host's connection to its network. Consequently, if a host machine is moved to another network, its address must be changed.

IP addresses are classified by their formats. Four formats are permitted: class A, class B, class C, and class D formats. As illustrated in Figure A–5, the first bits of the address specify the format of the remainder of the address field in relation to the network and host subfields. The host address is also called the local address (also called the REST field).

The *class A* addresses provide for networks that have a large number of hosts. The host ID field is 24 bits. Therefore, 2^{24} hosts can be identified. Seven bits are devoted to the network ID, which supports an identification scheme for as many as 127 networks (bit values of 1 to 127).

Class B addresses are used for networks of intermediate size. Fourteen bits are assigned for the network ID, and 16 bits are assigned for the

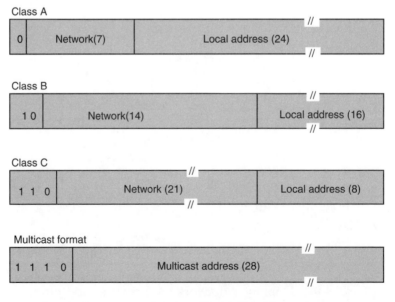

Figure A–5   Internet Protocol (IP) address formats.

host ID. *Class C* networks contain fewer than 256 hosts (2^8). Twenty-one bits are assigned to the network ID. Finally, *class D* addresses are reserved for multicasting, which is a form of broadcasting, but within a limited area.

The IP address space can take the forms as shown in Table A–2, and the maximum network and host addresses that are available for the class A, B, and C addresses are also shown.

There are instances when an organization has no need to connect into the Internet or another private intranet. Therefore, it is not necessary to adhere to the IP addressing registration conventions, and the organization can use the addresses it chooses. It is important to be certain that connections to other networks will not occur, since the use of addresses that are allocated elsewhere could create problems.

In RFC 1597, several IP addresses have been allocated for private addresses, and it is a good idea to use these addresses if an organization chooses not to register with the Internet. Systems are available that will translate private, unregistered addresses to public, registered addresses if connections to global systems are needed.

Table A–2   IP Addresses

	Network Address Space Values	
A	from: 0.0.0.0	to: 127.255.255.255*
B	from: 128 .0.0.0	to: 191.255.255.255
C	from: 192.0.0.0	to: 223.255.255.255
D	from: 224.0.0.0	to: 239.255.255.255
E	from: 240.0.0.0	to: 247.255.255.255**

* Numbers 0 and 127 are reserved
** Reserved for future use

	Maximum Network Numbers	Maximum Host Numbers
A	126 *	16,777,124
B	16,384	65,534
C	2,097,152	254

* Numbers 0 and 127 are reserved

The addresses set aside for private allocations:

Class A addresses: 10.x.x.x – 10.x.x.x (1)

Class B addresses: 172.16.x.x – 172.31.x.x (16)

Class C addresses: 192.168.0.x – 192.168.255.x (256)

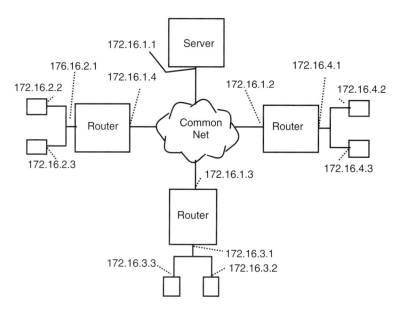

Figure A–6   Examples of IP addressing.

Figure A–6 shows examples of the assignment of IP addresses in more detail (examples use IP class B addresses). A common backbone (common net) connects three subnetworks: 176.16.2, 176.16.3, and 176.16.4. Routers act as the interworking units between the legacy (conventional) LANs and the backbone. The backbone could be a conventional Ethernet, but in most situations, the backbone is a Fiber Distributed Data Interface (FDDI), a Fast Ethernet node, or an ATM hub.

The routers are also configured as subnet nodes, and access servers are installed in the network to support address and naming information services.

The IP datagram contains the source address and the destination address of the sender and receiver, respectively. These two addresses do not change. They remain intact end-to-end. The destination address is used at each IP module to determine which "next node" is to receive the datagram. It is matched against the IP routing table to find the outgoing link to reach this next node.

In contrast, the MAC source and destination addresses change as the frame is sent across each link. After all, MAC addresses have significance only at the link layer.

In Figure A–7, the IP source address of A.1 and destination address of C.2 stay the same throughout the journey through the Internet. The MAC addresses change at each link. It is necessary for the destination MAC

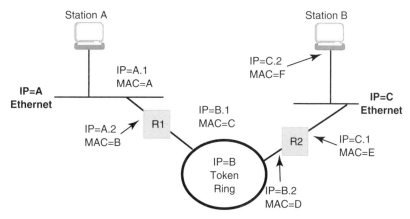

From	To	Source IP Address	Destination IP Address	Source MAC Address	Destination MAC Address
Station A	Router 1	A.1	C.2	A	B
Router 1	Router 2	A.1	C.2	C	D
Router 2	Station B	A.1	C.2	E	F

Figure A–7   Relationship of IP and MAC addresses.

address to contain the MAC address of the machine on the respective LAN that is to receive the frame. Otherwise, the frame cannot be delivered.

At first glance, it might appear that the IP addressing scheme is flexible enough to accommodate the identification of a sufficient number of networks and hosts to service almost any user or organization. But this is not the case. The Internet designers were not shortsighted; they simply failed to account for the explosive growth of the Internet and the rapid growth of the IPs in private networks.

The problem arises when a network administrator attempts to identify a large number of networks or computers (such as personal computers) attached to these networks. The problem becomes onerous because of the need to store and maintain many network addresses and the associated requirement to access these addresses through large routing tables. The use of address advertising to exchange routing information requires immense resources if they must access and maintain big addressing tables.

The problem is compounded when networks are added to an internet. The addition requires the reorganization of routing tables and perhaps the assignment of additional addresses to identify the new networks.

To deal with this problem, the Internet establishes a scheme whereby multiple networks are identified by one address entry in the routing table. Obviously, this approach reduces the number of network addresses needed in an internet. It also requires a modification to the routing algorithms, but the change is minor in comparison to the benefits derived.

Figure A–8 shows the structure of the slightly modified internet address. All that has taken place is the division of the local address, heretofore called the host address, into the subnet address and the host address.

It is evident that both the initial Internet address and the subnet address take advantage of hierarchical addressing and hierarchical routing. This concept fits well with the basic gateway functions inherent in the Internet.

The choice of the assignments of the "local address" is left to the individual network implementors. There are many choices in the definition of the local address. As we mentioned before, it is a local matter, but it does require considerable thought. It requires following the same theme of the overall Internet address of how many subnets must be identified in relation to how many hosts that reside on each subnet must be identified.

To support subnet addressing, the IP routing algorithm is modified to support a subnet mask. The purpose of the mask is to determine which

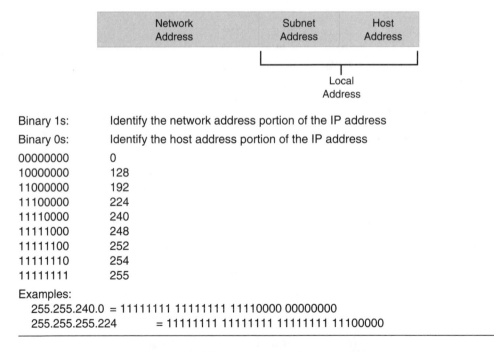

| Binary 1s: | Identify the network address portion of the IP address |
| Binary 0s: | Identify the host address portion of the IP address |

00000000	0
10000000	128
11000000	192
11100000	224
11110000	240
11111000	248
11111100	252
11111110	254
11111111	255

Examples:
    255.255.240.0 = 11111111 11111111 11110000 00000000
    255.255.255.224   = 11111111 11111111 11111111 11100000

Figure A–8   The subnet address structure.

part of the IP address pertains to the subnetwork and which part pertains to the host.

The convention used for subnet masking is to use a 32-bit field in addition to the IP address. The contents of the field (the mask) are set as shown in Figure A–8.

Table A–3 is provided to aid in correlating the IP binary subnet mask to hexadecimal and decimal equivalents. The table is self-descriptive.

Table A–4 is based on a table that appears in Chris Lewis's book.[1] It should be helpful if you are trying to determine how many subnets and hosts can be derived from different combinations of subnet masks. The tables are for class B and class C networks. One rule should be remembered: The first and last address of a host or subnet range of numbers cannot be used; they are reserved. So, if the range of the value is 3 bits (0–7), the values permit 6 addresses. The first bit identifies the actual subnet number, and the last bit is the broadcast address for that subnet. Here is an example from the Chris Lewis reference.

IP address	210.222.5.121
Subnet mask	255.255.255.248
Subnet address	201.222.5.120
Usable subnet host addresses	201.222.5.121–201.222.5.126
Subnet broadcast address	201.222.5.127

Table A–3   IP Subnet Mask Values

Binary Values				Hex Values				Decimal Values			
1111 1111	1111 1111	1111 1111	1111 1111	FF	FF	FF	FF	255	255	255	255
1111 1111	1111 1111	1111 1111	1111 1110	FF	FF	FF	FE	255	255	255	254
1111 1111	1111 1111	1111 1111	1111 1100	FF	FF	FF	FC	255	255	255	252
1111 1111	1111 1111	1111 1111	1111 1000	FF	FF	FF	F8	255	255	255	248
1111 1111	1111 1111	1111 1111	1111 0000	FF	FF	FF	F0	255	255	255	240
1111 1111	1111 1111	1111 1111	1110 0000	FF	FF	FF	E0	255	255	255	224
1111 1111	1111 1111	1111 1111	1100 0000	FF	FF	FF	C0	255	255	255	192

---

[1]The table in Chris Lewis's book on the class C subnet mask is in error (Mr. Lewis leaves out one iteration of 255; see page 38). It is not a big deal and does not detract from the overall quality of the book.

Table A–3   IP Subnet Mask Values *(continued)*

Binary Values				Hex Values				Decimal Values			
1111 1111	1111 1111	1111 1111	1000 0000	FF	FF	FF	80	255	255	255	128
1111 1111	1111 1111	1111 1111	0000 0000	FF	FF	FF	00	255	255	255	00
1111 1111	1111 1111	1111 1110	0000 0000	FF	FF	FE	00	255	255	254	00
1111 1111	1111 1111	1111 1100	0000 0000	FF	FF	FC	00	255	255	252	00
1111 1111	1111 1111	1111 1000	0000 0000	FF	FF	F8	00	255	255	248	00
1111 1111	1111 1111	1111 0000	0000 0000	FF	FF	F0	00	255	255	240	00
1111 1111	1111 1111	1110 0000	0000 0000	FF	FF	E0	00	255	255	224	00
1111 1111	1111 1111	1100 0000	0000 0000	FF	FF	C0	00	255	255	192	00
1111 1111	1111 1111	1000 0000	0000 0000	FF	FF	80	00	255	255	128	00
1111 1111	1111 1111	0000 0000	0000 0000	FF	FF	00	00	255	255	00	00
1111 1111	1111 1110	0000 0000	0000 0000	FF	FE	00	00	255	254	00	00
1111 1111	1111 1100	0000 0000	0000 0000	FF	FC	00	00	255	252	00	00
1111 1111	1111 1000	0000 0000	0000 0000	FF	F8	00	00	255	248	00	00
1111 1111	1111 0000	0000 0000	0000 0000	FF	F0	00	00	255	240	00	00
1111 1111	1110 0000	0000 0000	0000 0000	FF	E0	00	00	255	224	00	00
1111 1111	1100 0000	0000 0000	0000 0000	FF	C0	00	00	255	192	00	00
1111 1111	1000 0000	0000 0000	0000 0000	FF	80	00	00	255	128	00	00
1111 1111	0000 0000	0000 0000	0000 0000	FF	00	00	00	255	00	00	00
1111 1110	0000 0000	0000 0000	0000 0000	FE	00	00	00	254	00	00	00
1111 1100	0000 0000	0000 0000	0000 0000	FC	00	00	00	252	00	00	00
1111 1000	0000 0000	0000 0000	0000 0000	F8	00	00	00	248	00	00	00
1111 0000	0000 0000	0000 0000	0000 0000	F0	00	00	00	240	00	00	00
1110 0000	0000 0000	0000 0000	0000 0000	E0	00	00	00	224	00	00	00
1100 0000	0000 0000	0000 0000	0000 0000	C0	00	00	00	192	00	00	00
1000 0000	0000 0000	0000 0000	0000 0000	80	00	00	00	128	00	00	00
0000 0000	0000 0000	0000 0000	0000 0000	00	00	00	00	00	00	00	00

Table A–4   Class B and C Subnet Masks and Resultant Subnets and Hosts

Number of Bits	Subnet Mask	For Class B Resultant Subnets	Resultant Hosts
2	255.255.192.0	2	16392
3	255.255.224.0	6	8190
4	255.255.240.0	14	4094
5	255.255.248.0	30	2046
6	255.255.252.0	62	1022
7	255.255.254.0	126	510
8	255.255.225.0	254	254
9	255.255.225.128	510	126
10	255.255.225.192	1022	62
11	255.255.225.224	2046	30
12	255.255.225.240	4094	14
13	255.255.225.248	8190	6
14	255.255.225.252	16382	2

Number of Bits	Subnet Mask	For Class C Resultant Subnets	Resultant Hosts
2	255.255.225.192	2	62
3	255.255.225.224	6	30
4	255.255.225.240	14	14
5	255.255.225.248	30	6
6	255.255.225.252	62	2

ADDRESS AGGREGATION AND SUBNET MASKS AND PREFIXES

Address aggregation is introduced in Chapter 1. It is the method used today to reduce the size of the routing tables. It is quite similar to the use of subnet masks, with the following exceptions: (a) the net and subnet bits are contiguous and begin in the high-order (most significant) part of the address space, (b) a 32-bit submask is not used, rather (c) a prefix value is appended to the end of an address to describe how many bits are to be used as the mask. In most routers, addresses can be configured with a conventional decimal notation or a prefix.

Figure A–9 shows how address aggregation is used. The three subnets 172.16.1.0, 172.16.2.0, and 172.16.3.0 use a prefix of 24. This value means the net and subnet (actually, net and subnet lose their meaning

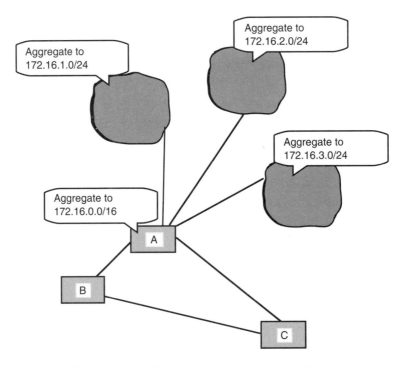

Figure A–9   Reducing routing table sizes.

now) span across the first 24 bits of the address, leaving the last 8 bits to identify the hosts.

However, router A does not have to advertise all three addresses. It aggregates these addresses into 172.16.0.0/16. The /16 means the first 16 bits of the advertised address pertain to networks (actually prefixes) at router A.

As a consequence of this approach, routers B and C do not have to store three addresses in their routing tables. They need only store one address with its prefix. Whenever routers B and C receive an IP datagram with 172.16.x.x in the destination address, the use of a stored prefix value in the routing table enables the routers to know that the datagram is to be sent to router A.

Routers B and C are not concerned with knowing about any more details of the bit contents of the address beyond the prefix. It is router A's job to know that the three networks are directly attached to router A's interfaces. Router A knows this fact because it has a special table containing the addresses of directly attached networks, and it consults this information before it accesses the long routing table.

Figure A–10 is a more detailed view of address aggregation. The arrows depict route-advertising packets, more commonly known as routing

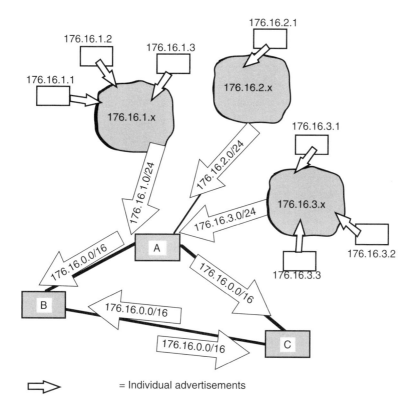

Figure A–10   The routing messages.

packets. The small arrows are advertisements being sent from hosts and are conveyed to an assigned router on each of the three subnets. These routers are not shown in the figure. In this simple example, seven hosts are sending routing packets. Each subnet is aggregated with the prefix of 24, resulting in three packets being sent to router A.

Router A aggregates these advertisements to a prefix of 16 and sends routing packets to its neighbor routers, B and C.

These routers send this same advertisement to each other. Consequently, routers B and C know of two routes to 176.16.0/16. Under most conditions, the most direct route would be chosen to these subnets; that is, directly through A. However, circumstances exist where B's packets to A might go through C first, and C's packets to A might go through B first. The obvious circumstance is a link failure between A and either B or C.

One other situation needs explaining in this example. Routers A, B, and C are connected in a loop. It is, therefore, conceptually possible for the routing packets to loop around over and over. Of course, measures are taken to preclude the looping of advertisements; these are explained in the "Loop Detection and Control" section in Chapter 5.

Figure A–11 shows how a subnet mask is interpreted. Assume a class B IP address of 128.1.17.1 with a mask of 255.255.240.0. At a router, to discover the subnet address value, the mask has a bitwise Boolean *and* operation performed on the address, as shown in the figure (this address is in a routing table). The mask is also applied to the destination address in the datagram.

The notation *"don't care,"* means that the router is not concerned at this point with the host address. It is concerned with getting the datagram to the proper subnetwork. So, in this example, it uses the mask to discover that the first 4 bits of the host address are to be used for the subnet address. Further, it discovers that the subnet address is 1.

As this example shows, when the subnet mask is split across octets, the results can be a bit confusing if you are "octet aligned." In this case, the actual value for the subnet address is 0001_2 or 1_{10}, even though the decimal address of the host space is 17.1. However, the software does not care about octet alignment. It is looking for a match of the destination address in the IP datagram to an address in a routing table, based on the mask that is stored in the routing table.

The class address scheme (A, B, C) has proved to be too inflexible to meet the needs of the growing Internet. For example, the class address of 47 means that three bytes are allocated to identify hosts attached to network 47, resulting in 2^{24} hosts on the single network—clearly not realistic! Moreover, the network.host address does not allow more than a two-

	128.	1.	17.	1
IP address	10000000	00000001	0001 0001	00000001
Mask	11111111	11111111	1111 0000	00000000
Result	10000000	00000001	0001	don't care
Logical address	128	1	1	don't care
	network		sub net	host

Note: "don't care" means router doesn't care at this time
(the router is looking for subnet matches)

Figure A–11   Example of address masking operations.

level hierarchical view of the address. Multiple levels of hierarchy are preferable, because they permit the use of fewer entries in routing tables and the aggregation of lower-level addresses to a higher-level address.

The introduction of subnets in the IP address opened the way to better utilize the IP address space by implementing a multilevel hierarchy. This approach allows the aggregation of networks to reduce the size of routing tables.

Figure A–12 is derived from [HALA 98] and shows the advertising operations that occur without route aggregation (without classless interdomain routing (CIDR), discussed next). The ISPs are ultimately advertising all their addresses to the Internet to a network access point (NAP). Four addresses are shown here, but in an actual situation, thousands of addresses might be advertised.

In contrast to the preceding example, where each address is advertised to the Internet, the use of masks allows fewer addresses to be

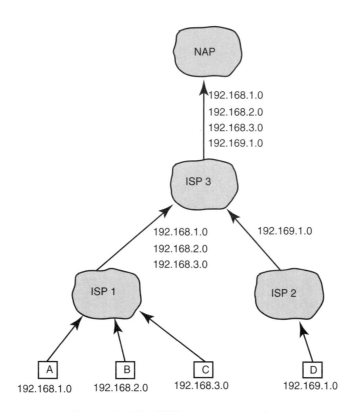

Figure A–12   Without aggregation.

advertised. In Figure A–13, ISP1 and ISP2 are using masks of 16 bits in length (255.255.0.0), and ISP1 need only advertise address 192.168.0.0 with the 16-bit mask to inform all interested nodes that all addresses behind this mask can be found at 192.168.x.x. ISP1 uses the same mask to achieve the same goal.

ISP3 uses a mask of 8 bits (255.0.0.0), which effectively aggregates the addresses of ISP1 and ISP2 under the aggregation domain of ISP3. Thus, in this simple example, one address instead of four is advertised to the NAP.

To extend the limited address space of an IP address, CIDR is now used in many systems and is required for operations between autonomous systems. It permits networks to be grouped logically and to use one entry in a routing table for multiple class C networks.

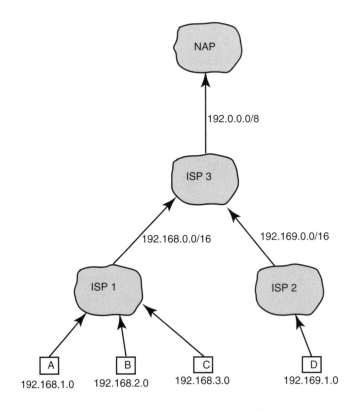

Note: The notations /16 and /8 refer to the lengths of masks.

Figure A–13   With aggregation.

Table A-5 Classless Interdomain Routing (CIDR) ("Supernetting")

- Reduces the size of routing tables
- Requirements:
  - Multiple IP addresses must share a specific number of high-order bits of an address
  - Masks must be used
  - Routing protocols must support the masks
- Example (from RFC 1466):
  - Addresses from 194.0.0.0 through 195.255.255.255
  - 65536 different class C addresses, but the first 7 bits are the same: 1100001x (they show the same high-order 7 bits)
  - One entry in a routing table of 194.0.0.0 with a mask of 254.0.0.0 suffices of all addresses (to a single point)
- A longer mask can be used to route to addresses beyond first mask

The example in Figure A-13 shows how the concept works. The first requirement for CIDR is for multiple networks to share a certain number of bits in the high-order part of the IP address. In the example, the first 7 bits in the address are the same. Thus, with the mask of 254.0.0.0 (11111110.00000000.00000000.00000000), all addresses between 194.0.0.0 and 195.255.255.255 can be identified by a single entry in the routing table.

Once that point in the network has been reached, the remainder of the address space can be used for hierarchical routing. For example, a mask of, say, 255.255.240.0 could be used to group networks. This concept, if carried out on all IP addresses (and not just class C addresses), would result in the reduction of an Internet routing table from about 10,000 entries to 200 entries.

Additional information on CIDR is available in RFCs 1518, 1519, 1466, and 1447, and summarized in Table A-5.

VARIABLE LENGTH SUBMASKS

Subnet masks are useful in internetworking operations, especially the variable-length subnet mask. Figure A-14 (which summarizes a more detailed example from [HALA98]) illustrates the idea of VLSM.

We assume an organization is using a class C address of 192.168.1.x. The organization needs to set up three networks (subnets) as shown in

Figure A–14. Subnet A has 100 hosts attached to it, and subnets B and C support 50 hosts each.

Recall from our previous discussions that the subnet mask is used to determine how many bits are set aside for the subnet and host addresses. Figure A–14 shows the possibilities for the class C address. (The resultant numbers in the table assume that the IP address reserved numbers

Class C Address of 192.168.1.x is used by an organization

Organization needs the following topology:

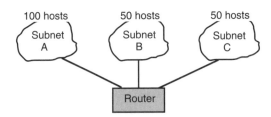

Possible Masks:

Subnet Mask	Resultant Subnets*	Resultant Hosts*
255.255.255.128	2	128
255.255.255.192	4	64
255.255.255.224	8	32
255.255.255.240	16	16
255.255.255.248	32	8
255.255.255.252	64	4

* Assumes use of reserved bits

Use .128 yields 2 subnets with 128 hosts each: Won't work

Use .192 yields 4 subnets with 64 hosts each: Won't work

Answer? Use both (Variable length subnet mask):

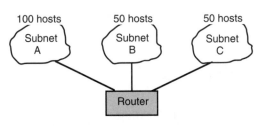

Subnet A mask = 255.255.255.128

Subnet B mask = 255.255.255.192

Subnet C mask = 255.255.255.192

Figure A–14   Managing the IP addresses.

are used, which is possible, since 192.168.1.x is from a pool of private addresses and can be used as the organization chooses.)

The use of one mask for the three subnets will not work. A mask of 255.255.255.128 yields only 2 subnets, and a mask of 255.255.255.192 yields only 64 hosts.

Fortunately, different subnetwork masks can be used on each subnet. As Figure A–14 shows, subnet A uses subnet mask 255.255.255.128, and subnets B and C use subnet mask 255.255.255.192.

Not all route discovery protocols support subnetwork masks. So check your product before you delve into this operation.

HIGH OVERHEAD OF IP FORWARDING

With the advent of subnet operations, the routing operations to support diverse topology and addressing needs are greatly enhanced. As an added bonus, the 32-bit IP address space is utilized more effectively.

However, these features translate into a more complex set of operations at the router. Moreover, as the Internet and internets continue to grow, the router may be required to maintain large routing tables. In a conventional routing operation, summarized in Figure A–14, the processing load to handle many addresses in combination with subnet operations can lead to serious utilization problems for the router.

Part of the overhead stems from the fact that a network can be configured with different subnet masks. The router must check each entry in the routing table to ascertain the mask, even though the table addresses may point to the same network. This concept, referred to as a variable-length submask (VLSM), provides a lot of flexibility in configuring different numbers of hosts to different numbers of subnets. For example, a class C address could use different subnet masks to identify different numbers of hosts attached to different subnets in an enterprise, say, 120 hosts at one site and 62 at another.

To route traffic efficiently, the router must prune (eliminate) table entries that do not match the masked portion of the table entry and the destination IP address in the datagram. After the table is pruned, the remaining entries must be searched for the longest match and more general route masks are discarded.

After all these operations, the router has to deal with the type of service (in practice, TOS may not be implemented), the best metric, and perhaps special procedures dealing with routing policies.

For high-end routers that are placed in the Internet to interwork between the large Internet service providers, the routing table could contain several thousand entries. To execute the operations of masking, table pruning, and longest mask matching requires extensive computational resources.

Because of this overhead, the high-end routers are taking a different approach, called label switching, the topic of this book.

Appendix B

CR-LDP and Traffic Engineering and QOS

In Chapters 5, 6, 7, and 9, the subjects of traffic engineering (TE) and quality of service (QOS) are explained in relation to MPLS. In several of these explanations, constraint-based routing with LDP (CR-LDP) was also included. As a follow-up to these discussions, this appendix provides more details on how CR-LDP interworks with DiffServ, ATM, RSVP, and Frame Relay to support the customer's QOS requirements. The tables in this appendix are sourced from a white paper by Nortel Networks, and I wish to thank Nortel Networks[1] for their contributions to my work on computer networks.

To understand the material in this appendix you should be familiar with material in the chapters cited above.

The tables in this appendix contain entries called *local behavior*. This term refers to the operations of the network nodes (typically label switching routes) in how they handle packet forwarding functions.

---

[1]*Using CR-LDP for Service Interworking, Traffic Engineering, and Quality of Service in Carrier Networks.* Document number 55049.25/09-00. Go to www.nortelnetworks.com or contact Marketing Publications, Dept. 9244, PO Box 13010, Research Triangle Park, NC 27709.

CR-LDP AND ATM QOS

In Chapter 6, I emphasized that MPLS must be able to interwork with ATM, since ATM is a prevalent technology in wide area and metropolitan networks. In Chapter 9, we saw several examples of the LDP message parameters, some of which are closely associated with ATM service categories. Table B–1 provides more details on these relationships.

For the ATM conformance definitions, the ITU-T and ATM Forum Generic Cell Rate Algorithm (GRCA) is used and is explained in Chapter 7. Many of the parameters listed in this table have been explained throughout this book, but it should prove helpful to find them in one place, with explanations on the ATM parameters (they are described in more detail in the ATM books in this series).

Parameters for ATM/CR-LDP Interworking

This part of the Appendix explains the ATM Forum's peak cell rate model and its relation to several CR-LDP traffic engineering operations.

The Peak Cell Rate Reference Model. The ATM Forum specifications provide a reference model to describe the peak cell rate (PCR). This model is shown in Figure B–1. It consists of an equivalent terminal, which contains the traffic sources, a multiplexer (MUX), and a virtual shaper. The term equivalent terminal means a model of a user device that performs the functions described in this discussion.

The traffic sources offer cells to a MUX, with each source offering cells at its own rate. Typically, the cells are offered from the AAL through the service access point (SAP). The MUX then offers all these cells to the virtual shaper. The virtual shaper smooths the cell flow that is offered to the physical layer and the ATM UNI (private UNI).

The GCRA comes into play in this model at three interfaces: (a) at the boundary between the ATM layer and the physical layer (the PhSAP, for physical service access point), (b) at the private user-to-network interface (UNI), and (c) at the public UNI.

For this discussion, we define T as the peak emission interval of the connection; the minimal interarrival time between two consecutive cells is greater than or equal to T. The PCR of the ATM connection is the inverse of the minimum arrival time between two cells.

The output of the virtual shaper at the PhSAP conforms to GCRA (T,0). The output at the private UNI and public UNI is different because

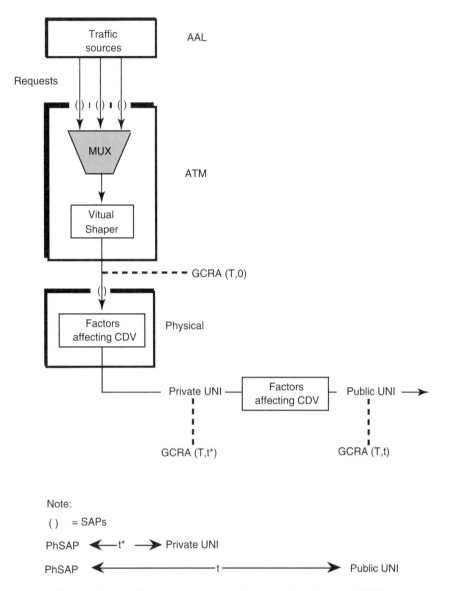

Figure B–1   Reference model for peak cell rate (PCR).

cell delay variation (CDV) will exist in the physical layer of the equivalent terminal (user device) and the node between this device and the network. Therefore, the private UNI conforms to GRCA (T,t^*), and the public UNI conforms to GRCA (T,t), where t is the cell delay tolerance. The latter value takes into consideration the additional CDV between the PhSAP and the public UNI.

The ATM network does not set the peak emission interval T at the user device. T can be set to account for different profiles of traffic, as long as the MUX buffers remain stable. Thus, T's reciprocal can be any value that is greater than the sustainable rate, but (of course) not greater than the link rate.

Cell Delay Variation Tolerance (CDVT). A certain amount of delay is encountered when cells are vying for the same output port of the multiplexer or when signaling cells are inserted in to the stream. As a result, with the reference to the peak emission interval T (which is the inverse of PCR R_p), randomness is instilled in the interarrival time between consecutive cells.

The 1-point CDV for cell k(y_k) at the measurement point is the difference between the cell's reference arrival time (c_k) and the actual arrival time (a_k) at the measurement point: $y_k = c_k - a_k$. The reference arrival time (c_k) is:

$$c_0 = a_0 = {}_0$$

$$c_k + 1 = \begin{cases} c_k + T \text{ if } c_k \geq a_k \text{ otherwise} \\ a_k + T \end{cases}$$

The 2-point CDV or cell k(v_k) between two measurement points MP_1 and MP_2 is the difference between the absolute cell transfer delay of cell k(x_k) between the two MPs and a defined reference cell transfer delay ($d_{1,2}$) between MP1 and MP_2: $v_k = v_k - d_{1,2}$.

The absolute cell transfer delay (xk) of cell k between MP_1 and MP_2 is the same as the cell transfer delay defined earlier. The reference cell transfer delay (d1,2) between MP_1 and MP_2 is the absolute cell transfer delay experienced by a reference cell between the two MPs.

CR-LDP Parameters

This part of the appendix reviews the CR-LDP parameters that were explained in several parts of this book.

- Peak data rate (PDR): The maximum rate at which the traffic can be sent to the CR-label switched path (CR-LSP). It is defined with a token bucket with the parameters peak data rate (PDR) and peak burst size (PBS).

- Peak burst size (PBS): The maximum burst size allowed at the peak data rate (PDR).
- Committed data rate (CDR): The rate that the MPLS domain commits for the CR-LSP. It is defined with the parameters committed data rate (CDR), and committed burst rate (CBR).
- Committed burst size (CBS): The maximum burst size allowed at the committed data rate (CDR). When the CBS bucket is full, it overflows into the excess burst rate (EBS) bucket.
- Excess burst size (EBS): Measures the extent to which traffic sent on the path exceeds the committed data rate (CDR). It is defined as an additional limit on the CDR/s token bucket.

CR-LDP Interworking with ATM QOS

Table B–1 shows the relationships of CR-LDP and ATM QOS. The label switching router (LSR) at the edge of the network is expected to enforce these parameters: (a) PDR, (b) PBS, (c) CDR, and (d) CBS.

CR-LDP AND FRAME RELAY QOS

Table B–2 shows the relationships of CR-LDP and Frame Relay services. Recall that Frame Relay basic operations are explained in Chapter 6. The Frame Relay service categories and parameters are explained as notes to Table B–2.

CR-LDP AND RSVP TRAFFIC ENGINEERING (RSVP-TE)

We learned in Chapter 5 that the Resource Reservation Protocol (RSVP) is used to reserve resources for a session in an Internet. RSVP can provide guaranteed service (GS) by reserving the necessary resources at each machine that participates in supporting the flow of traffic, as established in RFC 2212. Although not discussed in Chapter 5, RSVP is also defined to support a controlled load service (CLS), defined in RFC 2211. Table B–3 shows the relationships of CR-LDP and RSVP-TE.

Table B–1 Cross-Referencing of ATM Service Categories to CR-LDP QOS Parameters

ATM Service Categories		CR-LDP QoS Parameters		
ATM Services	ATM Conference Definition	Parameter Mapping	Edge Rules	Local Behavior
CBR	GCRA1(T0 + 1, CDVT)	PDR = 1/T0 + 1 PBS → CDVT	Drop packets > (PDR, PBS)	Very Frequent
VBR.1	GCRA1(T0 + 1 CDVT) GCRA2(Ts0 + 1 + BT 0 + 1 + CDVT)	PDR = 1/T0 + 1 PBS → CDVT CDR = 1/Ts0 + 1 CBS → BT0 + 1 + CDVT	Drop packets > (PDR, PBS) Drop packets > (CDR, CBS)	Frequent or Unspecified
VBR.2	GCRA1(T0 + 1, CDVT) GCRA2(Ts0 + BT0 + CDVT)	PDR = 1/T0 + 1 PBS → CDVT CDR = 1/Ts0 CBS → BT0 + CDVT	Drop packets > (PDR, PBS) Drop packets > (CDR, CBS)	Frequent or Unspecified
VBR.3	GCRA1(T0 + 1, CDVT) GCRA2(Ts0 + BT0 + CDVT)	PDR = 1/T0 + 1 PBS → CDVT CDR = 1/Ts0 CBS → BT0 + CDVT	Drop packets > (PDR, PBS) Tag packets > (CDR, CBS)	Frequent or Unspecified
UBR.1	GCRA(T0 + 1, CDVT)	PDR = 1/T0 + 1) PBS → CDVT	Network specific; (Tagging is not allowed.)	Unspecified
UBR.2	GCRA(T0 + 1, CDVT)	PDR = 1/T0 + 1 PBS → CDVT	Network specific; (Tagging is not allowed.)	Unspecified

Notes to Table B–1:

- The Frequency parameter — Frequent or Unspecified — for VBR services depends on whether or not the service is real-time.
- The mapping between the Cell Delay Variation Tolerance (CDVT) and the Peak Burst Size (PBS) and the Burst Tolerance (BT) and the Committed Data Size (CDS) is as specified in the ATM Forum standards.

Table B–2 Cross-Referencing of Frame Relay Services to CR-LDP QOS Parameters

| ATM Service Categories | | CR-LDP QOS Parameters | | |
Service	Parameters	Parameter Mapping	Edge Rules	Local Behavior
Default	AR, CIR, Bc, Be	PDR = AR CDR = CIR CBS = Bc EBS = Be	Police PDR, CDR, CBS, and EBS Drop Packets > PDR or > Be Tag packets > Bc but within Be	Unspecified
Mandatory	AR, CIR, Bc, Be	PDR = AR CDR = CIR CBS = Bc EBS = Be	Police PDR, CDR, CBS, and EBS Drop Packets > PDR or > Be Tag packet > Bc but within Be	Unspecified
Optional1	AR, CIR, Bc, Be	PDR = AR CDR = CIR CBS = Bc EBS = Be	Police PDR, CDR, CBS, and EBS Drop Packets > PDR or > Be Tag packet > Bc but within Be	Very Frequent
Optional2	AR, CIR, Bc, Be	PDR = AR CDR = CIR CBS = Bc EBS = Be	Police PDR, CDR, CBS, and EBS Drop Packets > PDR or > Be Tag packet > Bc but within Be	Frequent

Notes to Table B–2:

- Committed Rare Measurement Interval (Tc) — The time interval during which the user is allowed to send only the committed amount of data (Bc) and the excess amount of data (Bc). Tc is computed at Bc/CIR.
- Committed Information Rate (CIR) — The information transfer rate at which the network is committed to transfer under normal conditions. The rate is averaged over a minimum increment of time Tc. CIR is negotiated at call setup.
- Committed Burst Size (Bc) — The maximum committed amount of data a user may offer to the network during a time interval Tc. Bc is negotiated at call setup.
- Excess Burst Size (Be) — The maximum allowed amount of data by which a user can exceed Bc during a time interval Tc. Be is delivered, in general, with a lower than Bc. Be is negotiated at call setup.
- Access Rate (AR) — *Data rate of the user access channel.*

297

Table B-3   Cross-Referencing of RSVP-TE Services to CR-LDP QOS Parameters

RSVP-TE Service		CR-LDP QOS Parameters		
Service	Service Parameters	Parameter Mapping	Edge Rules	Local Behavior
GS	p, b, r, m, and M	PDR = p CDR = r CBS = b	• At any interval of length T, traffic should not exceed ([M + min[pT, rT, + b − M]). • Non-conforming packets are treated as best effort.	Very Frequent
CLS	b, r, m, and M p is optional	PDR = p CDR = r CBS = b	• At any interval of length T, traffic should not exceed [rT + b]. • Non-conforming packets are treated as best effort.	Frequent

Notes to Table B-3:

Similarities Between GS and CLS

• Both are real-time services that must be "hardwired" by correlating the edge rules to the local behavior. The main difference between the two services is the nature of the delay guarantee.

• Both services are defined at the network edge by the following token bucket parameters, which are collectively called Tspec parameters:
  - p = Peak rate of the flow (optional for CLS)
  - b = Bucket depth
  - t = Token bucket rate
  - m = Minimum policed unit
  - M = Maximum policed unit

• Traffic is policed at the edge, and the usual enforcing policy is to forward non-conforming packets as best effort.

Differences Between GS and CLS

• GS provides a hard mathematical upper boundary of packet delays, and CLS provides a delay that is equivalent to that seen by a best-effort service on a lightly loaded network.

• GS requires the reshaping of traffic to the token bucket parameters to meet the service delay requirements.

• GS also has Rspec parameters for the level of reservation in the RSVP domain, but because the RSVP terminates at the IWU and on Inter-working Unit, these parameters are not extended to the CR-LDP region.

References

[ASHW01] Ashwood-Smith, Peter et al. "Generalized MPLS-Signaling Functional Description, draft-ietf-mpls-generalized-signaling-05.txt, July 2001.

[AWDU99] Awduche, D., J. Malcolm, B. Agogbua, M. O'Dell, and J. McManus. "Requirements for Traffic Engineering Over MPLS," RFC 2702, September 1999.

[AWDU01] Awduche, Daniel, et al. "RSVP-TE: Extensions to RSVP for LSP Tunnels," RFC 3209.

[BERN99] Bernet, Y., A. Smith, and S. Blake. "A Conceptual Model for DiffServ Routers," draft-ietf-diffserv-model-00.txt, June 1999.

[BERN01] Bernstein, G., et al. "Optical Inter-Domain Routing Considerations," draft-bernstein-optical-bgp-01.txt, July 2001. NOTE: The text (but not the footnote) calls this reference [ONG01].

[BLAC00a] Black, Uyless. *IP Routing Protocols.* Prentice Hall, 2000.

[BLAC02] Black, Uyless. *Optical Networks: Third Generation Transport Systems.* Prentice Hall, 2002.

[CONT98] Conta, A., P. Doolan, and A. Malis. "Use of Label Switching on Frame Relay Networks," draft-ietf-mpls-fr-03.txt, November 1998.

[DAVI99] Davie, Bruce, et al.., "MPLS Using LDP and ATM VC Switching," draft-ietf-mpls-atm-02.txt, April 1999.

[FAUC00] Le Faucheur, Francois, Liwen Wu, and Bruce Davie. "MPLS Support of Differentiated Services," draft-ietf-mpls-diff-ext-04.txt, March 2000.

[GAN01] Gan, D., P. Pan, A. Ayyanger, and K. Kompella. "A Method for MPLS LSP Fast-Reroute Using RSVP Detours," draft-gan-fast-reroute-00.txt, April 2001.

[HALA98] Halabi, Bassam. Internet Routing Architectures. Cisco Press, 1998.

[HEIN99] Heinanen, J., et al. "Assured Forwarding PHB Group," RFC 2597, June 1999.

[JACO99] Jacobson, B., et al. RFC 2598, "An Expedited Forwarding (EF) PHB," June 1999.

[JAMO01] Jamoussi, Bilel, et al. "Constraint-Based LSP Setup Using LDP," draft-ietf-mpls-cr-ldp-05.txt, February 2001.

[KATZ01] Katz, Dave, et al. "Traffic Engineering Extension of OSPF," Draft-katz-yeung-ospf-traffic-05.txt. Expiration date: December 2001.

[KOMP02] Kompella, K., et al. "Extensions in Support of Generalized MPLS," draft-ietf-ccamp-ospf-gmpls-extensions-00.txt. Expiration date: March 2002

[MANN01] Mannie, E., et al. "Extensions to OSPF and IS-IS in support of MPLS for SDH/SONET Control," draft-mannie-mpls-sdh-ospf-isis-01.txt, July 2001.

[MART01] Martini, Luca, et al. "Encapsulation Methods for Transport of Layer 2 Frames Over MPLS," draft-martini-12-circuit-encap-mpls-01.txt, February 2001.

[NAGA01] Nagami, Ken-ichi, et al. "VCID Notification over ATM for LDP," RFC 3038, January 2001.

[NEWM98] Newman, Peter, Greg Minshall, and Tom Lyon. IP label switching—ATM Under IP," *IEEE / ACM Transactions on Communications* 6(2), April 1998.

[PART98] Partridge, Craig, Phil Carvey, E. Burgess, et al. "A 50-Gb/s Router," *IEEE ACM Transactions on Networking* 6(3), June 1998.

[ROSE01a] Rosen, Eric C. "MPLS Label Stack Encoding," RFC 3032, January 2001.

[ROSE01b] Rosen, Eric C., et al, BGP/MPLS VPNs, draft-ierf-ppvpn-rfc2547bis-00.txt, July 2001.

[ROSE01c] Rosen, Eric C., et al., "Multiprotocol Label Switching Architecture," RFC 3031, January 2001.

[ROSE01d] Rosen, Eric, et al. "OSPF as the PE/CE Protocol in BGP/MPLS VPNs," draft-rosen-vpns-ospf-bgp-mpls-03.txt, November 2001.

[WENT97] Wentworth, R., ATM Forum Contribution 97-0980, December 1997.

[WIDJ99] Widjaja, Il, and A. Elwalid. "Performance Issues in VC-Merge Capable ATM LSRs," RFC 2682, September 1999.

[WORS98] Worster, Tom, and Robert Wentworth. "Guaranteed Rate in Differentiated Services," draft-worster-diffserv-gr-00.txt, June 1998.

[XU01] Xu, Yangguang, et al. "A BGP/GMPS Solution for Inter-Domain Optical Networking," July 2001.

Acronyms

AAL	**ATM adaption layer**
ABR	available bit rate
ADM	add-drop multiplexer
AF	assured forwarding
ANSI	American National Standards Institute
AS	autonomous system
ASIC	application-specific integrated circuit
ATM	asynchronous transfer mode
BA	**behavior aggregate**
Bc	committed burst
Be	excess burst
BECN	backward explicit congestion notification (bit)
BGP	Border Gateway Protocol
BT	burst tolerance
BTT	bidirectional traffic trunk
CCITT	**Comite Consultatif International Téléphonique et Telegraphique**
CBS	committed burst size
CDR	committed data rate
CDV	cell delay variation
CDVT	cell delay variation tolerance
CE	customer edge (device)
CIDR	classless interdoman routing
CLP	cell loss priority (bit)
CLS	controlled load service
CoS	class of service
CPE	customer premises equipment
CQ	custom queuing

CR	constraint-based routing
CR-LDP	constraint-based LDP
CR-LSP	constraint-based routed label switched path
CSMA/CD	carrier-sense multiple access/collision detect
DE	**discard eligibility (indicator bit)**
DLCI	data link connection identifier
DNS	Domain Name System
DPA	destination physical address
DS	Differentiated Service
DSCP	DS codepoint
EBS	**excess burst size**
EF	expedited forwarding
EGP	Extended Gateway Protocol
EIGRP	Enhanced IGRP
ELAN LEC	emulated LAN configuration server
E-LSP	EXP-inferred-PSC LSP
ER	explicit routing
ESF	extended superframe
ETSI	European Telecommunications Standards Institute
EWMA	exponential weighted moving average
FCS	**frame check sequence (field)**
FDDI	Fiber Distributed Data Interface
FEC	forwarding equivalence class
FECN	forward explicit congestion notification (bit)
FDM	frequency division multiplexing
FIFO	first in, first out
FMP	Flow Management Protocol
FSC	fiber-switch capable
FTN	FEC-to-NHLFE
GC/PRA	**generic cell/packet rate algorithm**
GFC	generic flow control (field)
GMPLS	Generalized MPLS
GPID	generalized protocol ID
GPRA	generic packet rate algorithm
GR	guaranteed rate
GRCA	Generic Cell Rate Algorithm
GS	guaranteed service
GSMP	General Switch Management Protocol
HDLC	**High-Level Data Link Control**
HEC	header error control (field)
ICMP	**Internet Control Message Protocol**
IETF	Internet Engineering Task Force
IGMP	Internet Group Message Protocol
IGP	Internal Gateway Protocol
IGRP	Inter-Gateway Routing Protocol
ILM	incoming label map
IMPOA	Multiprotocol over ATM
IP	Internet Protocol

IPX	Internetwork Packet Exchange (protocol)
IS-IS	Intermediate System-to-Intermediate System (protocol)
ISP	Internet service provider
ISUP	ISDN user part
ITU-T	International Telecommunication Union-Telecommunication Standardization Bureau
L2TP	**Layer 2 Tunneling Protocol**
LAC	L2TP Access Concentrator
LAN	local area network
LANE	LAN emulation
LB	local binding
LCN	logical channel number
LCT	last conformance time
LDP	Label Distribution Protocol
LEC	local exchange carrier
LFIB	label forwarding information base
LIB	label information base
LLC	logical link connection
LMP	Link Management Protocol
LNS	L2TP Network server
LSA	link state advertisement
LSC	lambda-switch capable
L-LSP	label-inferred-PSC LSP
LSP	label switched panel (also: label switched path)
LSPID	label switched path ID
LSR	label switching router
MAC	**Media Access Control**
MAE	metropolitan area exchange
MF	multifield
MGR	multigigabit router
MPC	MPOA client
MPLS	Multiprotocol Label Switching
MPOA	Multiprotocol Over ATM
MPS	MPOA server
MUX	multiplexer
NAP	**network access point**
NAS	network access server
NBMA	nonbroadcast multiple access (network)
NHLFE	Next Hop Label Forwarding Entry
NHRP	Next Hop Resolution Protocol
NHS	next hop server
NNI	network-to-node interface
NSAP	Network Service Access Point (standard)
OA	**ordered aggregate**
OAM	operations, administration, maintenance,
OC	optical carrier
OSP	optical switched path
OSPF	Open Shortest Path First (protocol)

OSI	Open Systems Interconnection
OUI	organization unique identifier
OXC	optical cross-connect
PARC	**Palo Alto Research Center (Xerox)**
PBR	policy-based routing
PBS	peak burst size
PDR	peak data rate
PDH	plesiochronous digital hierarchy
PDU	protocol data unit
PDV	packet delay variation
PE	provider edge
PHB	per-hop behavior
PhSAP	physical service access point
PID	protocol ID
PNNI	Private Network-to-Network Interface
POS	packet on SONET
PPP	point-to-point protocol
PPS	packets per second
PPTP	Point-to-Point Tunneling Protocol
PQ	priority queuing
PSTN	public switch telephone network
PTI	payload type identifier (field)
PVC	permanent virtual circuit
PXC	packet-switched capable
QOS	**quality of service**
RD	**routing domain (also: route distinguisher)**
RESV	reservation
RFC	Request for Comments
RIP	Routing Information Protocol
RSVP	Resource Reservation Protocol
RTT	round-trip time
SAFI	**subsequent address family identifier**
SAP	service access point
SAR	segmentation and reassembly
SDH	synchronous digital hierarchy
SDU	service data unit
SONET	synchronous optical network
SLA	service level agreement
SNA	Systems Network Architecture
SP	service provider
SR	service representation
SS7	Signaling System Number 7
STDM	statistical time division multiplexing
STS	synchronous transport signal
SVC	switched virtual calls (also: switched virtual circuit)
TAT	**theoretical arrival time**
TB	token bucket
TCB	traffic conditioning block

TCP	Transmission Control Protocol
TDM	time division multiplexing (also: time division multiplex capable)
TDP	Tag Distribution Protocol
TE	traffic engineering
TFIB	tag forwarding information base
TIB	tag information base
TLV	type-length-value
TOS	type of service
TP	traffic profile
TSR	tag switching router
TTL	time to live
UDP	**User Datagram Protocol**
UNI	user-to-network interface
UPC	usage parameter control
VC	**virtual circuit**
VCC	virtual channel connection
VCI	virtual channel ID (also: virtual channel identifier)
VCID	virtual circuit ID
VLSM	variable-length submask
VoIP	Voice over IP
VP	virtual path
VPC	virtual path connection
VPN	virtual private networks
VPI	virtual path ID or identifier
VT	virtual tributary
WAN	**wide area network**
WDM	wave division multiplexing
WFIB	wavelength forwarding information base
WFQ	weighted fair queuing
ZIP (code)	**zoning improvement plan**

Index

Page numbers ending in "f" refer to figures. Page numbers ending in "t" refer to tables.